Biophysical Chemistry - Advance Applications

Edited by Mohammed A. A. Khalid

Published in London, United Kingdom

IntechOpen

Supporting open minds since 2005

Biophysical Chemistry - Advance Applications
http://dx.doi.org/10.5772/intechopen.73426
Edited by Mohammed A. A. Khalid

Part of IntechOpen Book Series: Physiology, Volume 7
Book Series Editor: Angel Catala

Contributors
Hidemi Toyoda, Dong Qing Xu, Lei Qi, Masahiro Hirayama, Saad Alosaimi, Awad Momen, Nikolay Lavrik, Manish Kumar, Yauheniya Osbon, Mohammed Khalid, Gabriela Ionita, Iulia Matei

First published in London, United Kingdom, 2020 by IntechOpen
IntechOpen is the global imprint of INTECHOPEN LIMITED, registered in England and Wales, registration number: 11086078, 7th floor, 10 Lower Thames Street, London, EC3R 6AF, United Kingdom
Printed in Croatia

British Library Cataloguing-in-Publication Data
A catalogue record for this book is available from the British Library

Additional hard and PDF copies can be obtained from orders@intechopen.com

Biophysical Chemistry - Advance Applications
Edited by Mohammed A. A. Khalid
p. cm.
Print ISBN 978-1-78984-047-6
Online ISBN 978-1-78984-048-3
eBook (PDF) ISBN 978-1-83880-138-0
ISSN 2631-8261

For EU product safety concerns:
IN TECH d.o.o., Prolaz Marije Krucifikse Kozulić 3, 51000 Rijeka, Croatia, info@intechopen.com or visit our website at intechopen.com.

IntechOpen Book Series
Physiology
Volume 7

Mohammed Khalid received his BSc degree in chemistry in July 2000, and his PhD degree in physical chemistry in 2007 from the University of Khartoum, Sudan. In 2009 he joined Dr. Ron Clarke's research group at the School of Chemistry, Faculty of Science, University of Sydney, Australia as a postdoctoral fellow where he worked on the interaction of ATP with the phosphoenzyme of the Na+,K+-ATPase and dual mechanisms of allosteric acceleration of the Na+, K+-ATPase by ATP. Thereafter he returned to the Department of Chemistry, University of Khartoum as an Assistant Professor and in 2014 he was promoted to Associate Professor. In 2011, he joined the staff of the Chemistry Department at Taif University, Saudi Arabia, where he is currently an Assistant Professor. His research interests include:

- P-type ATPase enzyme kinetics and mechanisms
- Kinetics and mechanism of redox reactions
- Autocatalytic reactions
- Computational enzyme kinetics
- Allosteric acceleration of P-type ATPases by ATP
- Exploring of allosteric sites of ATPases
- Interaction of ATP with ATPases in the cell membranes.

Editor of Volume 7: Mohammed A. A. Khalid
Department of Chemistry, Turabah University College, Taif University, Turabah, Saudi Arabia
Department of Chemistry, Faculty of Science, Khartoum University, Khartoum, Sudan

Book Series Editor: Angel Catala
National University of La Plata, Argentina

Scope of the Series

Modern physiology requires a comprehensive understanding of the integration of tissues and organs throughout the mammalian body, including the expression, structure, and function of molecular and cellular components. While a daunting task, learning is facilitated by our identification of common, effective signaling pathways employed by nature to sustain life. As a main example, the cellular interplay between intracellular Ca2 increases and changes in plasma membrane potential is integral to coordinating blood flow, governing the exocytosis of neurotransmitters and modulating genetic expression. Further, in this manner, understanding the systemic interplay between the cardiovascular and nervous systems has now become more important than ever as human populations age and mechanisms of cellular oxidative signaling are utilized for sustaining life. Altogether, physiological research enables our identification of clear and precise points of transition from health to development of multi-morbidity during the inevitable aging process (e.g., diabetes, hypertension, chronic kidney disease, heart failure, age-related macular degeneration; cancer). With consideration of all organ systems (e.g., brain, heart, lung, liver; gut, kidney, eye) and the interactions thereof, this Physiology Series will address aims of resolve (1) Aging physiology and progress of chronic diseases (2) Examination of key cellular pathways as they relate to calcium, oxidative stress, and electrical signaling & (3) how changes in plasma membrane produced by lipid peroxidation products affects aging physiology

Contents

Preface

Biophysical chemistry is one of the most interesting interdisciplinary research fields. Some of its different subjects have been intensively studied for decades. Now the field attracts not only scientists from chemistry, physics, and biology backgrounds but also those from medicine, pharmacy, and other sciences. We aimed to start this version of the book *Biophysical Chemistry* from advanced principles, as we include some of the most advanced subject matter, such as advanced topics in catalysis applications (first section) and therapeutic applications (second section). This led us to limit our selection to only chapters with high standards, therefore there are only six chapters, divided into two sections. We have assumed that the interested readers are familiar with the fundamentals of some advanced topics in mathematics such as integration, differentiation, and calculus and have some knowledge of organic and physical chemistry, biology, and pharmacy. We hope that the book will be valuable to graduate and postdoctoral students with the requisite background, and by some advanced researchers active in chemistry, biology, biochemistry, medicine, pharmacy, and other sciences.

I thank all chapter authors for their contributions to the book and the book processing manager Ms. Sara Debeuc for her valuable comments and assistance in the preparation of the book for publication. I hope that readers with an interest in biophysical chemistry will find the book interesting and of value to their own research.

Dr. Mohammed A. A. Khalid
Department of Chemistry,
Turabah University College,
Taif University,
Turabah, Saudi Arabia

Department of Chemistry,
Faculty of Science,
Khartoum University,
Khartoum, Sudan

Section 1

Catalytic Applications

Chapter 1

Introductory Chapter: The Diversity of Biophysical Chemistry Techniques

Mohammed Awad Ali Khalid

1. Introduction

Biophysical chemistry is an interdisciplinary field of study that uses concepts of chemistry and physics to understand biological systems by describing the quantitative, qualitative, energetics, structure, functions, and interactions phenomena of its physical nature. Because of the complexities of biological systems, a wide range of classical and sophisticated techniques have been employed in the field of biophysical chemistry; therefore mathematical, physical, and chemical techniques all flow into a single stream to describe biological systems. As an interesting field of research, biophysical chemistry is now growing up rapidly with major breakthroughs everywhere. In fact in 2009 Nobel Prize in chemistry have been awarded to three biophysical chemists for their work in X-ray diffraction of ribosomes [1]. The most important areas that attract biophysical chemists are molecular structure, molecular function, molecular dynamics, and kinetics, interactions, and thermodynamics of macromolecules that are located in the cell membrane or cytoplasmic constituents. In general biophysical chemistry interests in answering the following questions: how does a biological process take place, what types of molecules or particles are involved in this process and what are their structures, how long does it take for a biological process to take place and what are the energetics that accompany that change, what are the functions of biological molecules, and what are the consequences upon the cell if some biological molecules work disfunctionally?

In this introductory chapter, we would like to shed light into the importance of biophysical chemistry as a growing field of science and the broad diversity of techniques that have been used to elucidate the related phenomena. These techniques are of spectroscopic, electrochemical, thermal, and physiological origins, and we will not be able to cover all of these techniques in such an introductory chapter. Instead we will describe some selected techniques for studying the structure and functions of biological molecules. Some of these techniques adopt environmental and semi-environmental conditions as that for biological molecules in its native environment. We will focus on four types of techniques that up to this date are used in biophysical chemistry field. Each technique comes with a short description and is cited with appropriate reference for more details.

2. Thermal techniques

The most adequate and proper techniques that are used in this category are differential scanning calorimetry (DSC) and isothermal titration calorimetry (ITC),

IntechOpen

which provide unique complementary information for nucleic acids, modified nucleic acids, nucleic acid-ligand interactions, and protein-ligand interaction and are very useful methods in finding thermodynamic parameters using the same simple Gibbs equation. DSC is a direct, easy-to-use, model-independent measurement tool that can be used with various physicochemical methods to obtain structural and bonding information [2–4]. Originally DSC is used to study gaining (endothermic or heat absorption) or losing (exothermic or heat generation) heat upon the biological reactions or interactions as a function of temperature and time [5].

DSC technique measures a heat change during a temperature difference, which is radiated or absorbed by the sample, in a controlled way, based on a temperature difference between the sample and the reference material [6, 7].

When conducting a DSC experiment, the sample cell, which contains the molecules of interest dissolved in appropriate solvent, and the other cell called reference cell, which contains only the same amount of solvent, are heated simultaneously, and hence temperatures of both cells are raised identically over time. Depending on the type of the energy process, weather it is endothermic or exothermic, temperature is counted. In this regard the temperature of both cells is raised simultaneously until the difference is reached. If the process is heat gaining, then more heat is needed to equilibrate the two cells, meaning more energy is required to bring the sample to the same temperature as the reference; hence, the concept of heat excess comes into the picture.

New instruments allow setting a variety of experimental parameters such as number of scans, post scan temperature, scan range and rate, as well as feedback strength, and for the goodness of the results, slower scan rates provide higher resolution, while high feedback strength will give optimal interaction or reaction sensitivity. Because of the ability of the instrument to easily measure enthalpy changes, temperature difference, and phase transitions, it is now used in many applications, for example, studying protein interaction with ligands and drugs, protein mutations [8], protein folding [9–12], lipid interaction with drugs [13], protein interaction with lipids [14], DNA duplex stability, major and minor grooves in DNA and ligand binding, helix-coil transition in DNA, thermodynamics of DNA melting, as well as DNA-based binding interactions.

Another techniques is isothermal titration calorimetry. This technique is very sensitive in measuring energy released by interactions of molecule of interest and biological molecule. Mainly this technique is used for the qualitative and quantitative measurement of such interactions [15, 16]. With the aid of the Gibbs free energy equation and equation of equilibrium free energy, the interaction system can be easily understood. Two parameters can be derived directly from the ICT instrument which include the equilibrium constant for the binding process (K) and the binding stoichiometry (n).

3. Electrical techniques

This category illustrates the most advance and versatile techniques that have been used in biological systems. The success of electrochemical methods is obvious since experiments are adopted to study biological systems in vivo and in vitro. Both types need, for instance, very tiny electrodes that are capable of entering individual biological cell without damaging it. Therefore considerable efforts have been done for developing such electrode. Recent progress shows that a new generation of ultramicroelectrodes are in use. Studying single cells are constructed by preparing working electrodes generally from 5- to 10-μm diameter of carbon fibers. In this regard improving signal-to-noise ratio can be done considering that the electrode size should approach the size of the detection area of interest. On the other hand a greater number of electrochemical events can be detected by larger electrode sizes [17, 18].

To further narrow this application, a single biological molecule can be investigated using patch-clamp technique. This technique is capable of directly recording ionic current that could flow from a single ion channel. Moreover conductance and conformational changes between nonconducting and conducting conformations can be detected as well [19]. Since this technique could measure the conductance of a single channel, and as the conductance is the movement of ion through a specific area, this process could interfere with ion concentration. Therefore special considerations should be taken to avoid such interference.

The current recording from the instrument that is produced from a single or few channels can be used to derive two important phenomena: one is the conductance of a single channel, and the other is the time required for opening and closing the channel or what we call gating kinetics. Initially this can be achieved by inserting gramicidin A as model pores into planar lipid bilayer membranes. Using this step we look for sufficient amounts of ion channels to be inserted in the bilayer, and then both patch-clamp and voltage-clamp of expressed channels can be recorded, and background noise can be significantly reduced by applying gigaseal resistance, which could be achieved by forming such high resistance between recording electrode and the membrane patch. Patch-clamp methodology is reviewed extensively in the following references [20, 21].

Three techniques are nowadays applied to record single-channel conductance: the inside-out patch, the outside-out patch, and whole-cell patch. More details can be found in [22].

The keystone in the area of electrochemical techniques that are applied in biological systems is the appropriate electrode. The electrode that is actively used in biological analysis is now called bioelectrodes. By the name it should be microelectrodes or less since all biological operations taking place in the dimeter of that of living cells and more precisely its electrode whose tip diameter is less than 10 μm wile ultramicroelectrodes whose tip diameter is less than 1 μm. There are so many applications of such electrodes in biological systems. Among them are neurochemical analysis, mutagenicity and toxicity detection, analysis of blood ions and gases, blood flow analysis, and analysis of small molecules, nucleic acids, and proteins.

Some secondary techniques can be used in combination with patch-clamp recording to study specific proteins. For example, rapid perturbation techniques can be used to investigate pre-steady-state kinetics of membrane transporters. In this regard the current produced across the membrane by a transporter can be measured, and hence the mechanism behind can be elucidated. One problem with this is the feature of transporter itself since ion fluxes and consequently current produced by transporters are much lower than that produced by channels. Therefore whole-cell or inside-out patch method is used.

Another interesting technique that is actively used is voltage jump fluorometry combined with site-directed fluorescence labeling. This technique is mainly used to detect local protein motion in real time under native environmental conditions. Cysteine as amino acid is used for this purpose and could be inserted in specific location within a protein structure. The more useful way to do that is site-directed mutagenesis. Through this some amino acids can be cut and replaced by the desired ones, which eventually results in cysteine being in a desired location. New structure of protein can be expressed in *Xenopus* oocytes and allow for fluorescently active dyes to bind cysteine, for example, tetramethyl rhodamine-6-maleimide. Expressed protein can be stimulated by, for example, voltage pulse which leads eventually to recording of fluorescence signals and current responses in an electrophysiological experiment which allows structural changes to be correlated with specific ion transport steps, and hence detailed mechanism can be illustrated. More details can be found in the following references [23-26].

4. Spectroscopic techniques

Under this section lie so many advance techniques that successfully discover many features of biological systems. Few of them will be selected and discussed. The most important technique is X-ray crystallography [27], which is considered to be the primary tool for determining the macromolecule biological structure. This could be done by studying the X-ray scattering pattern that is produced from macromolecule prepared structure. Here we should note that not all biological macromolecules can form crystals. Some of them could form very nice crystals which have their molecules arranged in regular array called and this is used as a scattering surface of the crystal, biological molecules that aren't capable to form crystal could their structure be determined using solution NMR method, determination of the structure is subject to mathematical formulas of X-ray diffraction and scattering applied in the field, this process is very complicated and could take very long time to determine one macromolecular structure. Another application of X-ray is its use as a tool for spectroscopy. X-ray absorption spectroscopy (XAS) [28] is quite the same as the regular absorption methods, but this uses X-ray radiation for excitation of electrons. This technique is sensitive to the element that is involved in the absorption, and because of such sensitivity, this is become a growing technique. Two methods have been developed from XAS. The first one is to use XAS as X-ray absorption near edge structure (XANES), and hence element order, geometry, and oxidation state can be elucidated. The second one is to use XRA as extended X-ray absorption fine structure (EXAFS). In this case the active site by which metal ion can bind to macromolecule can be defined. The resolutions of both methods are in the range of Angstrom. One of the advantages of this methods is their ability to analyze any type of sample including perfect crystals, torsional crystals, and noncrystalline samples or amorphous samples, and the main disadvantage of this technique is its less sensitivity to the amount of material of interest that is less than mg/g.

Another interesting spectroscopic technique is stopped-flow fluorometry [24–26] in conjugation with voltage-sensitive fluorescence membrane probes. The stopped-flow technique allows the kinetics of ion pumps to be followed in the millisecond time range. Partial reactions of pumps which involve charge movement within the membrane produce local changes in electric field strength, which shift the fluorescence excitation spectrum of the probes and thus produce a fluorescence response.

Nuclear magnetic resonance (NMR) allows measurement of steady-state fluxes across cell membranes [29], which can be achieved due to the different chemical environment inside and outside cells. Signals coming from NMR active nuclei on both sides of cell membranes can be distinguished. Time-resolved changes in the intensity of these signals, thus, allow the steady-state kinetics of channels, pumps, and transporters to be detected and analyzed.

Time-resolved infrared spectroscopy recently became very frequent in use as a growing spectroscopic technique [30], since, with the help of pulsed lasers, it is possible to study processes that occur on timescales as short as 10-9 seconds or lower. Many transporters or pumps can be activated by the photochemical release of caged compounds. During ion transport processes, changes in the infrared absorbance spectrum of the pump or transporter under investigation can occur. Apart from following the kinetics of the ion transport reactions, in contrast to fluorescence spectroscopy, infrared spectroscopy allows more detailed structural information concerning the involvement of individual amino acid residues to be deduced.

Fluorescence correlation spectroscopy (FCS) is a technique with single-molecule sensitivity that correlates the fluorescence fluctuations due to the diffusion of individual fluorophores through the focal volume of the microscope [31, 32]. It can be used to quantify protein dynamics and concentrations, as well as protein–protein interactions in the dual-color version of the method. This technique is similar to single-particle tracking, which is a complementary technique that also provides information about protein dynamics and the oligomeric state of the proteins analyzed at the single-molecule level.

Another technique worth to be discussed is single-molecule fluorescence imaging [33]. In this case there is a wide range of single-molecule methods capable of examining transporter and channel and pump function in vivo and in vitro. With probing membrane protein molecules, conformational changes and measurements of ionic flux across the bilayer can be defined.

Flash photolysis is a very useful technique for studying the kinetics and mechanism of light-activated ion pumps and channels [34], which undergo changes in their UV/visible absorbance spectrum as part of their reaction cycle. One interesting example of kinetics and mechanism that have been resolved by this technique is that of the light-activated proton pump bacteriorhodopsin and the newly discovered light-activated channelrhodopsins.

Electron spin resonance as a powerful technique for structural elucidation of protein structure is extensively used [35, 36]. For this site-directed spin-labeling electron spin resonance spectroscopy (SDSL-EPR) is used as a highly sensitive technique for determining the structure and dynamics of, for example, membrane proteins within their native membrane environment. Site-directed mutagenesis methods are used to introduce a unique cysteine at a desired location to which a paramagnetic nitroxide spin label is then attached. The EPR spectrum yields information about side-chain mobility, solvent accessibility, the polarity of its immediate environment, and inter-nitroxide distances. Using these measurements it is possible to determine secondary structure, orientation of elements, and protein topography which allow questions concerning membrane protein structure, conformational changes, and membrane translocation phenomena to be investigated. The time-scale of sensitivity of the EPR technique is also well matched to the motions of lipids in membranes and conformational changes of functional relevance (picoseconds to seconds).

The last technique I would like to shed light on is atomic absorption spectrophotometry (AAS) [37]. It can be used to quantify the activity of cation transporting P-type ATPases by measuring cation uptake into individual cells by atomic absorption spectrophotometry using heterologous expression in *Xenopus* oocytes. The method is a sensitive and safe alternative to radioisotope flux experiments and is explicitly suited to the assessment of transport activity of electroneutrally operating transporters, which cannot easily be measured by electrophysiology. The method exhibits a low background of unspecific cation uptake and facilitates complex kinetic studies. Furthermore, since cation uptake can be determined on single cells, the flux experiments can be carried out in combination with two-electrode voltage clamping (TEVC) to achieve accurate control of the membrane potential and current.

5. Miscellaneous techniques

Long timescale molecular simulations [38–40]—In the past, the size and timescale of molecular simulations of macromolecules have been limited by the available computational power. As a consequence it has been difficult to include the full complexity of the natural environment of proteins and to simulate for long enough to capture biological processes taking longer than tens of nanoseconds. However, recently the steady increase in computer power has allowed for much longer simulations to be conducted, with a few cutting-edge studies reporting simulations of millisecond duration. Thus, it is now possible to directly investigate global conformational changes of proteins that are responsible for protein function as well as to quantify the energetic of physiological processes.

Radioactivity-based analysis technique [41]—Using this technique flux (influx and efflux) measurements of ions and substances across biological membranes using radioisotopes can be determined. Radioactive-based analysis of membrane transport is the method of choice if the transporters are operating electroneutrally (in this case the patch-clamp method cannot be applied) or if sensitive fluorescent dyes are not available. The latter is especially true for Na^+ and K^+ transport.

Author details

Mohammed Awad Ali Khalid[1,2]

1 Department of Chemistry, Turabah University College, Taif University, Turabah, Saudi Arabia

2 Department of Chemistry, Faculty of Science, Khartoum University, Khartoum, Sudan

*Address all correspondence to: m.khalid@tu.edu.sa; mkhalid@uofk.edu

References

[1] The Nobel Prize in Chemistry. 2009. Available from: https://www.nobelprize.org/prizes/chemistry/2009/summary/

[2] Chiu MH, Prenner EJ. Differential scanning calorimetry: An invaluable tool for a detailed thermodynamic characterization of macromolecules and their interactions. Journal of Pharmacy & Bioallied Sciences. 2011;**3**(1):39-59

[3] Bruylants G, Wouters J, Michaux C. Differential scanning calorimetry in life science: Thermodynamics, stability, molecular recognition and application in drug design. Current Medicinal Chemistry. 2005;**12**(17):2011-2020

[4] Fourches D. Application of computational techniques in pharmacy and medicine. Theoretical and Computational Chemistry. 2014;**17**:550

[5] Johnson CM. Differential scanning calorimetry as a tool for protein folding and stability. Archives of Biochemistry and Biophysics. 2013;**531**(1-2):100-109

[6] Haynie DT. Biological Thermodynamics. Cambridge, UK: Cambridge University Press; 2008

[7] Haines PJ, Reading M, Wilburn FW. Differential thermal analysis and differential scanning calorimetry. In: Brown ME, editor. Handbook of Thermal Analysis and Calorimetry. Vol. 1. The Netherlands: Elsevier Science BV; 1998. pp. 279-361

[8] Sturtevant JM. Effects of mutations on thermodynamic properties of proteins. ACS Symposium Series. 1993;**516**:1-17

[9] Holdgate GA, Ward WH. Measurements of binding thermodynamics in drug discovery. Drug Discovery Today. 2005;**10**:1543-1550

[10] Minetti CA, Remeta DP. Energetics of membrane protein folding and stability. Archives of Biochemistry and Biophysics. 2006;**453**:32-53

[11] Freire E. Thermal denaturation methods in the study of protein folding. In: Energetics of Biological Macromolecules. Vol. 259. San Diego: Academic Press Inc; 1995. pp. 144-168

[12] Eftink MR. Use of multiple spectroscopic methods to monitor equilibrium unfolding of proteins. In: Energetics of Biological Macromolecules. Vol. 259. San Diego: Academic Press Inc; 1995. pp. 487-512

[13] Lúcio M, Lima JL, Reis S. Drug-membrane interactions: Significance for medicinal chemistry. Current Medicinal Chemistry. 2010;**17**:1795-1809

[14] Mouritsen OG, Bloom M. Models of lipid-protein interactions in membranes. Annual Review of Biophysics and Biomolecular Structure. 1993;**22**:145-171

[15] Happi Emaga T et al. Purification of pectin from apple pomace juice by using sodium caseinate and characterization of their binding by isothermal titration calorimetry. Food Hydrocolloids. 2012;**29**:211-218

[16] Ognjenović J et al. Interactions of epigallo-catechin 3-gallate and ovalbumin, the major allergen of egg white. Food Chemistry. 2014;**164**:36-43

[17] Cahill PS, Walker QD, Finnegan JM, Mickelson GE, Travis ER, Wightman RM. Microelectrodes for the measurement of catecholamines in biological systems. Analytical Chemistry. 1996;**68**:3180-3186

[18] Travis ER, Wightman RM. Spatio-temporal resolution of exocytosis from individual cells. Annual Review of

Biophysics and Biomolecular Structure. 1998;**27**:77-103

[19] Neher E, Sakmann B. Single-channel currents recorded from membrane of denervated frog muscle fibres. Nature. 1976;**260**:799-802

[20] Ogden DC, editor. Microelectrode Techniques: The Plymouth Workshop Handbook. 2nd ed. Plymouth: The Company of Biologists; 1994

[21] Walz W, editor. Patch-Clamp Analysis: Advanced Techniques. 2nd ed. Totowa, NJ: Humana Press; 2007

[22] Andrew JM, Trevor ML, Peter HB. Analyzing ion permeation in channels and pumps using patch-clamp recording. In: Clarke RJ, Khalid MAA, editors. Pumps, Channels, and Transporters: Methods of Functional Analysis. first ed. 2015. pp. 51-88

[23] Sven G, Jack HK, Ernst B, Thomas F. Conformational dynamics of the Na,K-ATPase probed by voltage clamp fluorometry. PNAS. 2003;**100**(3):964-969

[24] Clarke R, Khalid M. Pumps, Channels and Transporters: Methods of Functional Analysis. USA: John Wiley & Sons; 2015

[25] Khalid M, Fouassier G, Apell H-J, Cornelius F, Clarke R. Interaction of ATP with the phosphoenzyme of the Na,K-ATPase. Biochemistry. 2010;**49**:1248-1258

[26] Khalid M, Cornelius F, Clarke R. Dual mechanisms of allosteric acceleration of the Na,K-ATPase by ATP. Biophysical Journal. 2010;**98**(10):2290-2298

[27] Gordon GH. Spectroscopy for the Biological Sciences. USA: John Wiley & Sons, Inc.; 2005

[28] Richard O, Asuncion C, Isabelle L. Pier LS. X-ray absorption spectroscopy of biological samples. A tutorial. Journal of Analytical Atomic Spectrometry. 2012;**27**:2054

[29] Fan TW-M, Lane AN. Applications of NMR spectroscopy to systems biochemistry. Progress in Nuclear Magnetic Resonance Spectroscopy. 2016;**92-93**:18-53

[30] Tanumoy M, Soma B, Subrata B, Pal SK. Study of biomolecular recognition using time-resolved optical spectroscopy. International Review of Biophysical Chemistry. 2011;**2**(6):211-237

[31] Radek M, Thorsten W. Recent applications of fluorescence correlation spectroscopy in live systems. FEBS Letters. 2014;**588**:3571-3584

[32] Perevoshchikova IV, Kotova EA, Antonenko YN. Fluorescence correlation spectroscopy in biology, chemistry, and medicine. Biochemistry (Moscow). 2011;**76**:497

[33] Jordanka Z, van Kensal H. Single-molecule biology: What is it and how does it work. Molecular Cell. 2006;**24**:317-329

[34] Dewey H, Maurice WW. Picosecond flash photolysis in biology and biophysics. Annual Review of Biophysics and Bioengineering. 1978;**7**:189-227

[35] Wertz JE, Bolton JR. Biological applications of electron spin resonance. In: Electron Spin Resonance. Dordrecht: Springer; 1986. pp. 378-390

[36] Shin CS, Dunnam CR, Borbat PP, Dzikovski B, Barth ED, Halpern HJ, et al. ESR microscopy for biological and biomedical applications. Nanoscience and Nanotechnology Letters. 2011;**3**(4):561-567

[37] Decker WJ. Atomic absorption spectroscopy: Applications in

agriculture, biology, and medicine. Archives of Internal Medicine. 1971;**128**(4):649-650

[38] Klepeis JL, Lindorff-Larsen K, Dror RO, Shaw DE. Long-timescale molecular dynamics simulations of protein structure and function. Current Opinion in Structural Biology. 2009;**19**(2):120-127

[39] Elber R. Long-timescale simulation methods. Current Opinion in Structural Biology. 2005;**15**(2):151-156

[40] Dodson GG, Lane DP, Verma CS. Molecular simulations of protein dynamics: New windows on mechanisms in biology. EMBO Reports. 2008;**9**(2):144-150

[41] Reisz JA, Bansal N, Qian J, Zhao W, Furdui CM. Effects of ionizing radiation on biological molecules-mechanisms of damage and emerging methods of detection. Antioxidants & Redox Signaling. 2014;**21**(2):260-292

Chapter 2

Application of Riboflavin Photochemical Properties in Hydrogel Synthesis

Gabriela Ionita and Iulia Matei

Abstract

Riboflavin, known as vitamin B2, belongs to the class of water-soluble vitamins with redox, fluorescence, and photosensitizing properties. Riboflavin contains a fragment of 7,8-dimethyl-10-(1′-D-ribityl) isoalloxazine with a system of conjugated double bonds that is are responsible for its photochemical properties. In the presence of light and oxygen, riboflavin generates reactive oxygen species that can be further involved in the oxidation of biological molecules such as amino acids, proteins, nucleotides, and lipids. The chapter focuses on the photochemical application of riboflavin in (1) cross-linking of structural proteins such as collagen and (2) synthesis of hydrogels. The involvement of riboflavin in such processes has already found application in medicine, especially in the treatment of ophthalmic diseases and in tissue engineering.

Keywords: riboflavin, hydrogel, cross-linking, extracellular matrix, polysaccharides

1. Introduction

Riboflavin (**Figure 1**), known as vitamin B2, has been intensively investigated since it was discovered and separated, initially from cow milk by Blyth and, almost 50 years later, from brewers' yeast. Riboflavin belongs to the flavin class and its structure (7,8-dimethyl-10-[(2S,3S,4R)-2,3,4,5-tetrahydroxypentyl]benzo[g]pteridine-2,4-dione) has been proven by chemical synthesis by Kuhn and Karrer. It consists of a ribityl side chain attached to an isoalloxazine ring with a system of conjugated double bonds that is responsible for the photochemical properties of riboflavin [1–4].

As a vitamin, riboflavin fulfills a number of essential biological functions, being involved in redox and photoreactions with nucleic acids, in the destruction of tumor cells and inactivation of viruses or bacteria [5, 6]. Riboflavin is a structural component of two flavoproteins with coenzyme role, flavin mononucleotide (FMN) and flavin adenine dinucleotide (FAD) (**Figure 1**). These take part in various and complex enzymatic processes such as the synthesis, conversion, and recycling of vitamins B1, B6, and folate and the synthesis of heme proteins, in fatty acid metabolism and in the regulation of thyroid hormones [7–9]. Riboflavin is thus involved in the citric acid and methionine cycles. Moreover, riboflavin and its metabolites have antioxidant properties and participate in the glutathione redox cycle [7, 10].

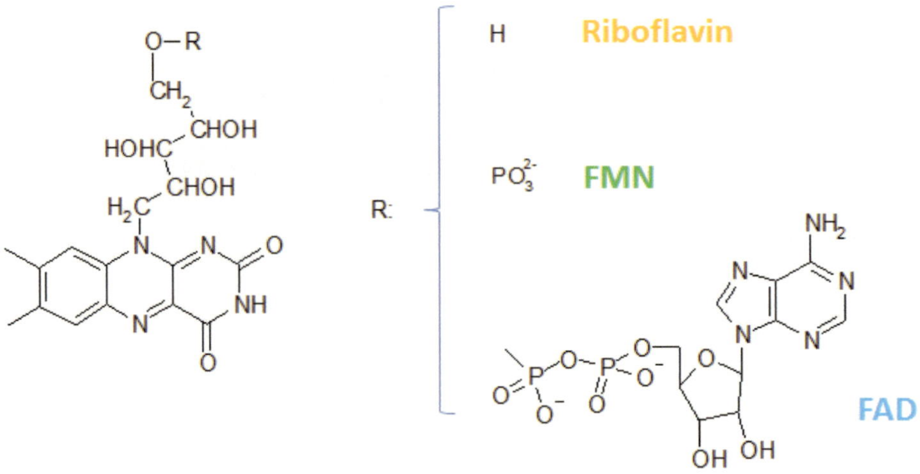

Figure 1.
Structures of riboflavin and its derivatives flavin mononucleotide (FMN) and flavin adenine dinucleotide (FAD).

Obviously, the involvement of riboflavin in biochemical processes is due to the flavin core. In aqueous solution at pH 7, the UV-Vis absorption spectrum of riboflavin is characterized by the presence of four bands with maxima at 223, 267, 373, and 444 nm [11, 12]. All bands have high molar extinction coefficients (> 10^4 M^{-1} cm^{-1}) that are indicative of π-π^* electronic transitions. The band positions and molar extinction coefficients are sensitive to the particularities of the environment sensed by the flavin chromophore. In water at pH 7, riboflavin emits yellow green fluorescence at 520 nm [13]. The fluorescence quantum yield and lifetime of the first excited singlet state S_1 are 0.26 [14] and 5 ns [15, 16], respectively. These values are influenced by the polarity and protic/aprotic character of the solvent, as well as by the pH value. The decrease in solvent polarity determines a hypsochromic shift of the emission band and an increase in the fluorescence quantum yield [17, 18], and this has been exploited to study different chemical systems formed with cyclodextrins and serum albumins [19–21]. Although neutral riboflavin presents strong fluorescence, the protonated species forming at pH < 4 are nonfluorescent, while the anionic species at pH > 9.7 are weakly fluorescent [22]. Flavoenzymes containing FMN as prosthetic group have similar emission intensity with riboflavin, while those containing FAD show a drastic drop in emission caused by the presence of adenosine [23].

Riboflavin undergoes various reactions under thermal or chemical treatment and under exposure to UV light [11]. Under irradiation, riboflavin can undergo photolysis to various degradation products and can produce reactive oxygen species that may further alter the structure and biological activity of proteins, lipids, and, to a lesser extent, carbohydrates [24]. Significant for this review are the processes that can be controlled by the reactive radical species formed in a system in the presence of riboflavin and UV light.

Photophysical and photochemical studies involving riboflavin are important for understanding the biological processes taking place in the presence of light and mediated by flavin derivatives and their further applications. In the presence of light, riboflavin is excited to a singlet state of higher energy, followed by nonradiative intersystem crossing to an excited triplet state. From this state, riboflavin can be involved in a photosensitized oxidation process following one of two mechanisms schematically presented in **Figure 2**. In type I mechanism, riboflavin transfers the energy to a substrate and generates riboflavin free radical species that further

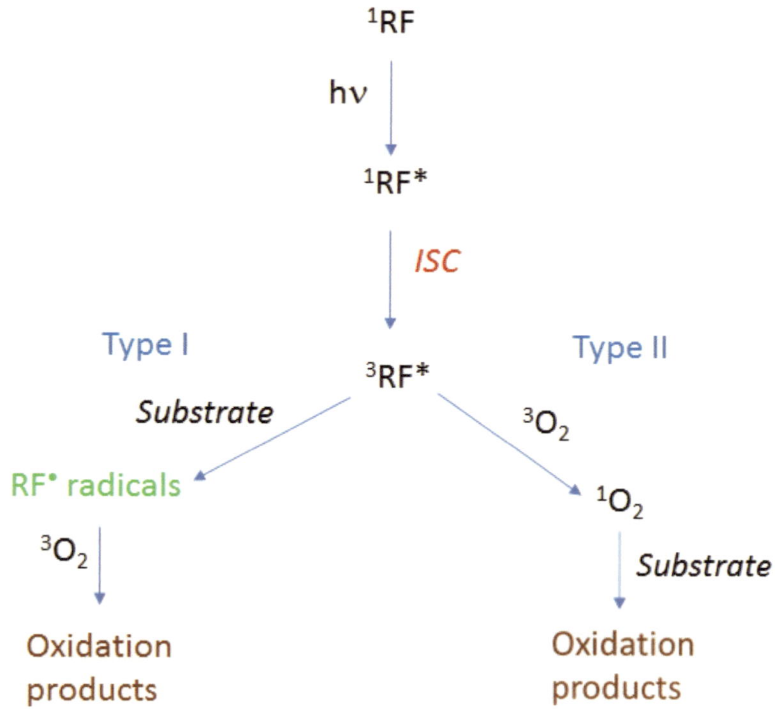

Figure 2.
Scheme of the photosensitized oxidation mechanisms mediated by riboflavin.

interact with molecular oxygen in the ground state to yield oxidation products. This mechanism occurs preferentially at low oxygen concentration. In type II mechanism, riboflavin transfers the energy to molecular oxygen in the ground state to generate the more reactive singlet molecular oxygen. The latter, in reaction with a substrate, leads to the final oxidation products [25].

The property of riboflavin to be involved in biological redox processes is currently exploited for the photodegradation of water pollutants including pesticides, herbicides, phenol, toluene, and toluene derivatives [26, 27]. Another research field that is expanding in what concerns the application of the photochemical properties of riboflavin regards the cross-linking process of the components of the extracellular matrix (ECM). Photoinduced collagen cross-linking is nowadays a method applied for the treatment of various diseases that involve components of the ECM, such as varicose veins [28], degraded dentine [29–31], and a series of ophthalmic diseases (keratoconus, pellucid marginal corneal degeneration, post-LASIK ectasia, infectious keratitis, bullous keratopathy) [32–35].

2. Collagen cross-linking in the presence of riboflavin and UVA light

The photochemical properties of riboflavin in the presence of UVA light are nowadays applied in the treatment of some ocular diseases, including keratoconus, in order to induce cross-linking of collagen that improves the biomechanical stability of corneal stroma and increases the resistance to enzymatic digestion [32–34]. The mechanism of corneal collagen cross-linking involves excitation of riboflavin to the triplet state, followed by generation of reactive oxygen species—$HO^{\bullet}$, $O_2^{\bullet 2-}$, and $O_2^{\bullet -}$ in the case of type I mechanism—or production of singlet oxygen ($^1O_2$) that can further generate peroxide and hydroxyl radicals, in the case of type II mechanism [12, 36, 37]. The schematic representation of this process is shown in **Figure 3**. In fact, both mechanisms depicted in **Figure 2** lead to the formation of reactive oxygen species. Formation of such reactive species favors reactions with carbonyl groups from the protein chain, which further increase the protein cross-linking. The production of reactive radical species can be evidenced by electron paramagnetic resonance (EPR) spectroscopy, in spin trapping experiments.

There are some studies reported in literature regarding the radical species formed in systems that contain collagen in the presence of riboflavin and UV light. In water solution of riboflavin, an environment exposed to oxygen, the formation of hydroxyl radicals was evidenced under the influence of UV radiation. This radical was also present in a solution of collagen and riboflavin [38]. Later, another EPR study was reported, aiming to monitor the formation of radical species generated in the presence of riboflavin in solutions of H_2O_2 and catalase or bovine serum albumin [39]. It was found that hydroxyl radicals are formed both in the presence of catalase and BSA, as the DMPO-OH adducts gave the characteristic signal with four lines in the intensity ratio of 1:2:2:1. The stable radical species such as TEMPO (2,2,6,6-tetramethylpiperidine 1-oxyls) derivatives can report on the changes in the tear protein compositions during the treatment of dry eye syndrome that is often associated with keratoconus disease [40]. Obviously, during the UVA-riboflavin collagen cross-linking treatment, the tears collected from patients contain reactive radical species. In a preliminary EPR study aiming to evidence radical species formed in solutions containing proteins that are present in a significant amount in tears, the spin trapping experiments evidenced the presence of the DMPO-OH adduct (the lines are marked with red dots in **Figure 4**) but also a carbon-centered radical. This second component of the spectrum (the corresponding lines are marked with blue dots in **Figure 4**) demonstrates that some other radicals can form during riboflavin/UVA treatment.

The EPR studies clearly demonstrate a radical mechanism of the processes occurring in the ECM protein components in the presence of riboflavin, mainly referring to collagen cross-linking. Changes in the collagen structure of human sclera tissues induced by riboflavin-UVA irradiation were evidenced by Raman spectroscopy and atomic force microscopy (AFM) [41]. Raman spectroscopy revealed changes in the position of some bands associated with specific vibration modes. The most significant changes were observed for the following vibration modes: (CCO), (CCC) stretch, and (S-S) stretch. Another feature observed in the Raman spectra of human sclera tissue after riboflavin/UVA irradiation treatment was the increase in the intensity and sharpness of the bands. After the sclera collagen cross-linking occurred, the AFM images revealed arrangements of interlocked

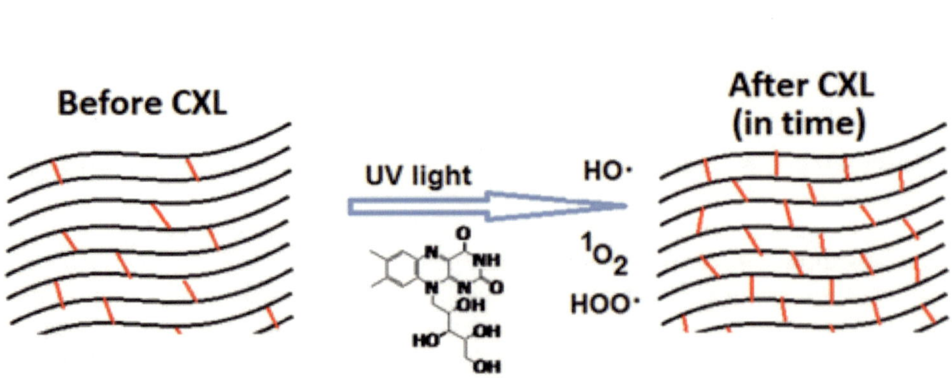

Figure 3.
Schematic representation of the collagen cross-linking process in the presence of riboflavin and UVA light.

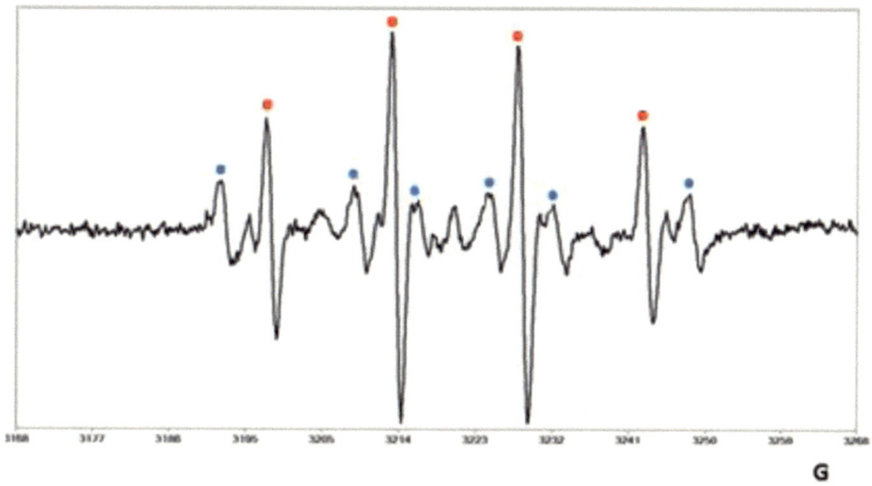

Figure 4.
DMPO adducts of HO$^{\bullet}$ (red) and carbon-centered (blue) radicals formed in collagen solution exposed to UVA light in the presence of riboflavin.

collagen fibrils. Both Raman spectra and AFM images proved the formation of interfibrillar cross-links in sclera tissues.

The ECM is responsible for the enhancement of cell proliferation, differentiation, and cell-to-cell interaction. Therefore, the information obtained on the structure and properties of collagen in natural tissues can be used as reference for building new networks that can mimic the ECM and can find application in tissue engineering, with implantology as target. Many studies revealed that the self-assembled collagen artificial networks are mechanically weaker compared to natural collagen in vivo [42]. However, some studies reported that riboflavin cross-linking improves the physicochemical properties of collagen scaffolds [43]. The collagen matrix built in vivo under the action of riboflavin and UV light has properties similar to those of natural collagen and is able to retain water for a longer time.

The collagen fibril formation is favored by the presence of glycosaminoglycans such as chondroitin sulfate and hyaluronic acid, both of them being parents of the natural ECM with a common function to control tissue hydration and to act as lubricants. Therefore, it is not surprising that these polysaccharide polymers are used to generate various hydrogels that can be used in medicinal applications.

3. Synthesis of hydrogels with medicinal applications mediated by riboflavin

Presently, the photochemistry of riboflavin can be exploited to control the physicochemical properties of new gel materials generated by the assembly of polymers that are similar to those found in the natural ECM. These novel materials are dedicated to medicinal applications such as tissue engineering involving space-filling agents, as vehicles for bioactive molecule delivery, for cell encapsulation, and wound healing.

Hydrogels can be classified according to the processes involved in the generation of their solid network. Supramolecular hydrogels result either through non-covalent assembly of most often small molecular weight gelators, while polymeric hydrogels are obtained through covalent polymer cross-linking. Supramolecular gels are more dynamic and sensitive to physical factors that generate their network, while polymeric structures have a more robust frame. The biomedical applications of hydrogels depend on their biocompatibility and physicochemical properties. For instance, it is expected for supramolecular hydrogels to find applications in drug delivery, while polymeric hydrogels can be used in tissue engineering. Synthetic polymers like poly(ethylene oxide) (PEO), poly(vinyl alcohol) (PVA), or poly(acrylic acid) (PAA) can be used in this specific biochemical application, although the naturally derived polymers present in the ECM—hyaluronic acid, chondroitin sulfate, collagen, and gelatin, are more suitable because the resulting materials show a better mimicking of the natural tissues [44].

The number of reports on the supramolecular gels resulted by assembly of low-molecular gelators involving riboflavin is relatively small. Formation of this type of gels often involves the supramolecular assembly of two gelators [45]. Saha et al. [46, 47] reported data on hydrogels that have riboflavin as gelator. They reported in a communication the assembly of riboflavin with melanin into a gel network [46], and the mechanistic aspects of this supramolecular process have been discussed later in a full paper [47]. Both molecules can be involved in multiple hydrogen bonds and π-π interactions of their aromatic moieties, which are in fact the pillars of the gel network formation. The ribityl chain of riboflavin ensures the water solubility of the riboflavin-melanin complex due to the hydroxyl groups present in the structure [46]. Polar solvents like water and hydrogen bond formation decrease the photophysical parameters of riboflavin [17, 18, 48]. Surprisingly, Saha et al. found that in the temperature interval corresponding to the gel phase of riboflavin/melanin, the photoluminescence spectra show increased intensity [46]. Electron microscopy techniques revealed that the gel fibrils of these gel systems have a helical pattern, which sustains the initial hypothesis of π-π stacking interactions. These results were corroborated with FT-IR, circular dichroism, and photoluminescence data [47]. Other hydrogels built on riboflavin and aromatic structures with functional groups that can be involved in hydrogen bonding, such as uric acid [49], gallic acid [50], quinazolinedione derivatives [51], salicylic acid, dihydroxy benzoic acid, and acetoguanamine [52], have been reported. These studies regarding different pairs of gelators revealed that the gel properties depend on the complementarity between gelators.

The studies on polymeric hydrogels that contain riboflavin or are generated making use of the photochemical properties of riboflavin are more numerous. The ability of riboflavin to generate peroxide radicals in the presence of UV light and oxygen represents a nontoxic way to initiate acrylamide- or methacrylate-based hydrogel networks [53, 54]. Batchelor et al. [55] have reported the formation of a hydrogel through the reaction of multi-arm polyethylene glycol, functionalized with norbornene moieties, with dithiothreitol (DTT) as cross-linker, in the presence of riboflavin excited by blue light. In this way, they have generated the thiyl radical that reacts rapidly with the double bond from norbornene to generate a carbon-centered radical. Following the propagation scheme, this carbon-centered radical reacts with another thiol group. The gelation process is dependent on the initiator concentration and light intensity [55].

The spectral properties of riboflavin depend on the pH, and this feature has been used by Patra et al. to monitor the release of two drugs, ciprofloxacin and metronidazole, from a biocompatible hydrogel obtained by cross-linking poly (methacrylic acid) with glycogen that was further labeled with riboflavin. The resulted hydrogel was characterized by physicochemical methods that indicated a reversible swelling behavior and pH-dependent drug release properties [56].

Wank et al. [57] investigated the recognition interactions in an interpenetrating polymer network based on poly(acrylic acid) and poly(vinyl alcohol). This hydrogel has been imprinted with 1-(4-methoxyphenyl)-5-methyl-1,2,3-triazol-4-carboxylic acid (MMTCA), and its adsorption properties have been determined. The ability of this gel network to load MMTCA has been compared with those to uptake of aspirin and riboflavin. Competitive loading from mixed solutions has shown that MMTCA is selectively uptake relative to aspirin and riboflavin. The results have been explained by analyzing different structural factors including the ability of hydroxyl and carboxyl groups to form hydrogen bonds with the gel network, as well as stereo-shape aspects of the uploaded compounds.

Glycol chitosan functionalized with methacrylate groups can generate a hydrogel network in the presence of riboflavin under exposure to visible blue light [58, 59]. The gelation can occur in situ; therefore, this system can be considered for applications in tissue engineering. The chitosan hydrogel can incorporate chondroitin and type II collagen—two components of the cartilaginous ECM that can themselves be used as scaffolds for tissue engineering. In this way, the instability and rapid enzymatic degradation of these ECM components are reduced. This chitosan-based hydrogel supports the growth of chondrocytes [58], and the incorporation of collagen and chondroitin sulfate increases chondrogenesis and favors cellular condensation [59]. The chitosan hydrogel described is stable if incubated at body temperature for 42 days in the absence of lysozyme but is degraded in the presence of lysozyme at concentrations that are found in human cartilage (up to 3 mg/g) [59]. Chondroitin and collagen II are retained into the hydrogel during the incubation time, although initially a small quantity has been released. All these results recommend this chitosan derivative hydrogel as a biomaterial that can find biomedical applications in cartilage regeneration.

Functionalization of natural polysaccharides with high occurrence in different tissues can generate biomaterials that have a stronger adhesion to body tissues. In this sense, Donnelly et al. [60] have reported the photo-hydrogelation of tyramine-derivatized hyaluronate by UV exposure in the presence of various concentrations of riboflavin. The network of hydrogel is the result of cross-linking processes following the reactions depicted in **Figure 5**. Functionalization of hyaluronate with aromatic residues of tyramine allows a hydrophobic interaction with the tyrosine residues of collagen from ECM. Although the material is not suitable for cell encapsulation, it can be used for various other applications like filling small fissure

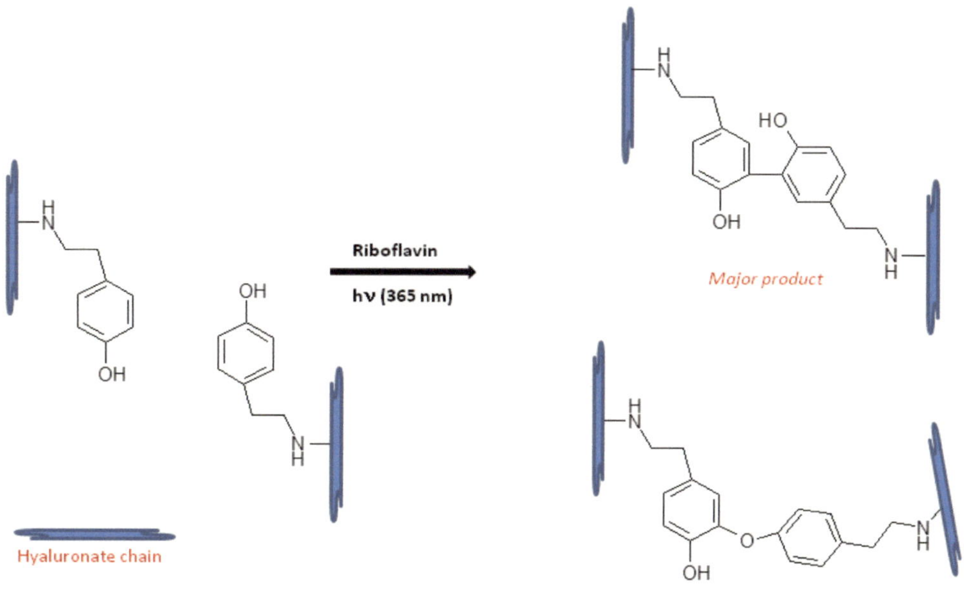

Figure 5.
Schematic representation of photo-cross-linking of TA-polymer in the presence of riboflavin.

defects, ensuring an interface between osteochondral grafts and host tissues in order to prevent edema, or acting as glue between the graft and the tissue.

The chitosan/riboflavin pair can be used in filling demineralized dentin substrates, due to the ability of riboflavin to generate, under UV light, radicals that initiate the cross-linking of collagen fibrils from dentin and chitosan. Fawzy et al. [61] have studied in detail the changes in dentin morphology by treating demineralized dentin with chitosan solutions in various concentrations, in the presence of riboflavin, over a period of 6 months. Their results showed improved mechanical properties of dentin and an enhanced stability of the collagen network.

Hyaluronic acid, as biocompatible polymer, is often used in drug formulations. Derivatives of hyaluronic acid such as methacrylate hyaluronic acid and thiolated hyaluronic acid are also biocompatible. In the presence of riboflavin and blue light (458 nm), slow gelation occurs [62], a process that has a different kinetic compared to the system described in Ref. 45. The induced gelation in a solution of these two hyaluronic derivatives occurs in 15 min at body temperature, and this delay in reaching the gel state can be exploited for in situ applications in ophthalmology and stomatology. The systems were characterized by rheological methods and FT-IR, and the ability to encapsulate and release bovine serum albumin and their cyto-compatibility were tested. Interestingly, in the absence of thiolated hyaluronic acid, the gelation process does not occur, while in the absence of riboflavin and light, the transition from sol to gel occurs even slower (in 24 h), and the resulting gel is weaker (the gel moduli have lower values). The gelation process is the result of two reactions: the oxidation of thiol groups leading to the formation of disulfide bonds (evidenced by IR spectroscopy) and the addition of the thiol group to the double bond of the methacrylate group.

A similar gel was obtained by reaction of methacrylated dextran with L-arginine in the presence of riboflavin, L-arginine initiating the photo-cross-linking [63]. The authors found that the riboflavin concentration influences the rate of gelation, an optimum value being in the range 0.2–1.0 wt%. At low riboflavin concentration, the resulting materials are brittle, while at higher concentration, they are compliable. The pH also affects the gelation process, a neutral pH being more favorable. The presence of arginine was essential as cross-linker, since gelation did not occur in its absence. The concentration of arginine was found to influence the properties of the gel. Thus, the weight ratio of arginine/methacrylate precursor in the range 0.8–1.2 led to gels with desirable properties in terms of shape stability, stickiness, and compliability. Microscopic images (SEM) indicated the honeycomb structure of the hydrogel in the swollen state.

4. Conclusions

In conclusion, in this review, we presented the general pathways for the generation of reactive species in the presence of riboflavin and UV-Vis light. The use of riboflavin as photoinitiator for polymerization processes presents the advantage of it being nontoxic and biocompatible. The research papers mentioned in this review also highlight the potential medicinal applications of gels formed in the presence of riboflavin.

Acknowledgements

This work was supported by the Romanian National Authority for Scientific Research, CNCS–UEFISCDI, project numbers PN-III-P4-IDPCE-2016-0734 and PN-III-P2-2.1-PED-2016-0187.

Conflict of interest

The authors declare no conflict of interest.

Author details

Gabriela Ionita* and Iulia Matei
Romanian Academy, "Ilie Murgulescu" Institute of Physical Chemistry, Bucharest, Romania

*Address all correspondence to: gabi2ionita@yahoo.com

References

[1] Northrop-Clewes CA, Thurnham DI. The discovery and characterization of riboflavin. Annals of Nutrition and Metabolism. 2012;**61**:224-230. DOI: 10.1159/000343111

[2] Bacher A, Eberhardt S, Eisenreich W, Fischer M, Herz S, Illarionov B, et al. Biosynthesis of riboflavin. Vitamins and Hormones. 2001;**61**:1-49. DOI: 10.1016/S0083-6729(01)61001-X

[3] Revuelta JL, Ledesma-Amaro R, Lozano-Martinez P, Díaz-Fernández D, Buey RM, Jiménez A. Bioproduction of riboflavin: A bright yellow history. Journal of Industrial Microbiology and Biotechnology. 2017;**44**(4-5):659-665. DOI: 10.1007/s10295-016-1842-7

[4] Schwechheimer SK, Park EY, Revuelta JL, Becker J, Wittmann C. Biotechnology of riboflavin. Applied Microbiology and Biotechnology. 2016;**100**(5):2107-2119. DOI: 10.1007/s00253-015-7256-z

[5] Reddy HL, Dayan AD, Cavagnaro J, Gad S, Li J, Goodrich RP. Toxicity testing of a novel riboflavin-based technology for pathogen reduction and white blood cell inactivation. Transfusion Medicine Reviews. 2008;**22**:133-153. DOI: 10.1016/j.tmrv.2007.12.003

[6] Insińska-Rak M, Sikorski M. Riboflavin interactions with oxygen—A survey from the photochemical perspective. Chemistry--A European Journal. 2014;**20**:15280-15291. DOI: 10.1002/chem.201403895

[7] Kennedy DO. B Vitamins and the brain: Mechanisms, dose and efficacy—A review. Nutrients. 2016;**8**:68-96. DOI: 10.3390/nu8020068

[8] Edwards AM. General properties of flavins. In: Silva E, Edwards AM, editors. Flavins Photochemistry and Photobiology. Cambridge: The Royal Society of Chemistry; 2006. pp. 1-11. DOI: 10.1039/9781847555397-00001

[9] Massey V. Chemical and biological versatility of riboflavin. Biochemical Society Transactions. 2000;**28**:283-296. DOI: 10.1042/bst0280283

[10] Ashoori M, Saedisomeolia A. Riboflavin (vitamin B2) and oxidative stress: A review. The British Journal of Nutrition. 2014;**111**:1985-1991. DOI: 10.1017/S0007114514000178

[11] Sheraz MA, Hafeez Kazi S, Ahmed S, Anwar Z, Ahmad I. Photo, thermal and chemical degradation of riboflavin. Beilstein Journal of Organic Chemistry. 2014;**10**:1999-2012. DOI: 10.3762/bjoc.10.208

[12] Beztsinna N, Sole M, Taib N, Bestel I. Bioengineered riboflavin in nanotechnology. Biomaterials. 2016;**80**:121-133. DOI: 10.1016/j.biomaterials.2015.11.050

[13] Weber G. Fluorescence of riboflavin and flavin adenine dinucleotide. The Biochemical Journal. 1950;**47**:114-121. DOI: 10.1042/bj0470114

[14] Koziol J. Studies on flavins in organic solvents-I. Spectral characteristics of riboflavin, riboflavin tetrabutyrate and lumichrome. Photochemistry and Photobiology. 1966;**5**:41-54. DOI: 10.1111/j.1751-1097.1966.tb05759.x

[15] Sikorska E, Khmelinskii IV, Komasa A, Koput J, Ferreira LFV, Herance JR, et al. Spectroscopy and photophysics of flavin related compounds: Riboflavin and iso-(6,7)-riboflavin. Chemical Physics. 2005;**314**:239-247. DOI: 10.1016/j.chemphys.2005.03.005

[16] Wahl P, Auchet JC, Visser AJWG, Müller F. Time resolved fluorescence of flavin adenine dinucleotide. FEBS Letters. 1974;**44**:67-70. DOI: 10.1016/0014-5793(74)80307-8

[17] Grodowski MS, Veyret B, Weiss K. Photochemistry of flavins-II. Photophysical properties of alloxazines and isoalloxazines. Photochemistry and Photobiology. 1977;**26**:341-352. DOI: 10.1111/j.1751-1097.1977.tb07495.x

[18] Naman SA, Tegnér L. Decay kinetics of the triplet excited state of lumiflavin. Photochemistry and Photobiology. 1986;**43**:331-333. DOI: 10.1111/j.1751-1097.1986.tb05612.x

[19] Morrison PW, Connon CJ, Khutoryanskiy VV. Cyclodextrin-mediated enhancement of riboflavin solubility and corneal permeability. Molecular Pharmaceutics. 2013;**10**: 756-762. DOI: 10.1021/mp3005963

[20] Neacsu MV, Ionita G, Topala C, Oprea E, Tecuceanu V, Matei I. Poly(ethylene glycol)/β-cyclodextrin covalent gel networks: Host matrices for studying radical processes in plant extract-riboflavin systems following UV irradiation. Chemical Papers. 2017;**71**:607-616. DOI: 10.1007/s11696-016-0047-x

[21] Voicescu M, Angelescu DG, Ionescu S, Teodorescu VS. Spectroscopic analysis of the riboflavin - Serum albumins interaction on silver nanoparticles. Journal of Nanoparticle Research. 2013;**15**:1555. DOI: 10.1007/s11051-013-1555-z

[22] Drossler P, Holzer W, Penzkofer A, Hegemann P. pH dependence of the absorption and emission behaviour of riboflavin in aqueous solution. Chemical Physics. 2002;**282**:429-439. DOI: 10.1016/S0301-0104(02)00731-0

[23] Heelis PF. The photophysical and photochemical properties of flavins (isoalloxazines). Chemical Society Reviews. 1982;**11**:15-39. DOI: 10.1039/CS9821100015

[24] Ahmad I, Vaid FHM. Photochemistry of flavins in aqueous and organic solvents. In: Silva E, Edwards AM, editors. Flavins Photochemistry and Photobiology. Cambridge: The Royal Society of Chemistry; 2016. Chapter 2. DOI: 10.1039/9781847555397-00013

[25] Choe E, Huang R, Min DB. Chemical reactions and stability of riboflavin in foods. Journal of Food Science. 2005;**70**:R28-R36. DOI: 10.1111/j.1365-2621.2005.tb09055.x

[26] Cui H, Hwang HM, Zeng K, Glover H, Yu HT, Liu YM. Riboflavin-photosensitized degradation of atrazine in a freshwater environment. Chemosphere. 2002;**47**:991-999. DOI: 10.1016/S0045-6535(02)00006-1

[27] Haggi E, Bertolotti S, Miskoski S, Amat-Guerri F, Garcia N. Environmental photodegradation of pyrimidine fungicides. Kinetics of the visible-light-promoted interactions between riboflavin and 2-amino-4-hydroxy-6-methylpyrimidine. Canadian Journal of Chemistry. 2002;**80**:62-67. DOI: 10.1139/v01-192

[28] Frullini A, Manetti L, Di Cicco E, Fortuna D. Photoinduced collagen cross-linking: A new approach to venous insufficiency. Dermatologic Surgery. 2011;**37**:1113-1118. DOI: 10.1111/j.1524-4725.2011.02047.x

[29] Cai J, Palamara JEA, Burrow MF. Effects of collagen crosslinkers on dentine: A literature review. Calcified Tissue International. 2018;**102**:265-279. DOI: 10.1007/s00223-017-0343-7

[30] Fawzy AS, Nitisusanta LI, Iqbal K, Daood U, Neo J. Riboflavin as a dentin

crosslinking agent: Ultraviolet A versus blue light. Dental Materials. 2012;**28**:1284-1291. DOI: 10.1016/j.dental.2012.09.009

[31] Liu XQ, Zhou JF, Chen L, Yang Y, Tan JG. UVA-activated riboflavin improves the strength of human dentin. Journal of Oral Science. 2015;**57**:229-234. DOI: 10.2334/josnusd.57.229

[32] Spoerl E, Huhle M, Seiler T. Induction of cross-links in corneal tissue. Experimental Eye Research. 1998;**66**:97-103. DOI: 10.1006/exer.1997.0410

[33] Wollensak G, Iomdina E, Dittert DD, Salamatina O, Stoltenburg G. Cross-linking of scleral collagen in the rabbit using riboflavin and UVA. Acta Ophthalmologica Scandinavica. 2005;**83**:477-482. DOI: 10.1111/j.1600-0420.2005.00447.x

[34] Spoerl E, Wollensak G, Seiler T. Increased resistance of crosslinked cornea against enzymatic digestion. Current Eye Research. 2004;**29**:35-40. DOI: 10.1080/02713680490513182

[35] Hsu KM, Sugar J. Keratoconus and other corneal diseases: Pharmacologic cross-linking and future therapy. Pharmacologic Therapy of Ocular Disease. 2017;**242**:137-161. DOI: 10.1007/164_2016_23

[36] Silva E, Jopia M, Edwards AM, Lemp E, Lissi E. Protective effect of Boldo and tea infusions on the visible light–mediated pro-oxidant effects of vitamin B2, riboflavin. Photochemistry and Photobiology. 2002;**75**:585-590. DOI: 10.1562/0031-8655(2002)075<0585:PEOBAT>2.0.CO;2

[37] Besaratinia A, Kim SI, Bates SE, Pfeifer GP. Riboflavin activated by ultraviolet A1 irradiation induces oxidative DNA damage-mediated mutations inhibited by vitamin C. Proceedings of the National Academy of Sciences of the United States of America. 2007;**104**:5953-5958. DOI: 10.1073/pnas.0610534104

[38] Constantin MM, Corbu C, Ionita G. EPR study on the role of riboflavin used in photo-oxidative collagen cross-linking. Revista de Chimie. 2010;**61**:495-497

[39] Sel S, Nass N, Pötzsch S, Trau S, Simm A, Kalinski T, et al. UVA irradiation of riboflavin generates oxygen-dependent hydroxyl radicals. Redox Report. 2014;**19**:72-79. DOI: 10.1179/1351000213Y.0000000076

[40] Constantin MM, Corbu CG, Tanase C, Codrici E, Mihai S, Popescu ID, et al. Spin probe method of electron paramagnetic resonance spectroscopy – a qualitative test for measuring the evolution of dry eye syndrome under treatment. Analytical Methods. 2019;**11**:965-972. DOI: 10.1039/C8AY02783H

[41] Jung GB, Lee HJ, Kim JH, Lim JI, Choi S, Jin KH, et al. Effect of cross-linking with riboflavin and ultraviolet A on the chemical bonds and ultrastructure of human sclera. Journal of Biomedical Optics. 2011;**16**:125004. DOI: 10.1117/1.3662458

[42] Rich H, Odlyha M, Cheema U, Mudera V, Bozec L. Effects of photochemical riboflavin-mediated crosslinks on the physical properties of collagen constructs and fibrils. Journal of Materials Science. Materials in Medicine. 2014;**25**:11-21. DOI: 10.1007/s10856-013-5038-7

[43] Chan BP, So KF. Photochemical crosslinking improves the physicochemical properties of collagen scaffolds. Journal of Biomedical Materials Research. Part A. 2005;**75**:689-701. DOI: 10.1002/jbm.a.30469

[44] Drury JL, Mooney DJ. Hydrogels for tissue engineering: Scaffold design variables and applications. Biomaterials. 2003;**24**:4337-4351. DOI: 10.1016/S0142-9612(03)00340-5

[45] Hirst AR, Smith DK. Two-component gel-phase materials—highly tunable self-assembling systems. Chemistry--A European Journal. 2005;**11**:5496-5508. DOI: 10.1002/chem.200500241

[46] Manna S, Saha A, Nandi AK. A two component thermoreversible hydrogel of riboflavin and melamine: Enhancement of photoluminescence in the gel form. Chemical Communications. 2006;**41**:4285-4287. DOI: 10.1039/b608234c

[47] Saha A, Manna S, Nandi AK. A mechanistic approach on the self-organization of the two-component thermoreversible hydrogel of riboflavin and melamine. Langmuir. 2007;**23**:13126-13135. DOI: 10.1021/la7026219

[48] Penzer GR, Radda GK. The chemistry and biological function of isoalloxazines (flavines). Quarterly Reviews, Chemical Society. 1967;**21**:43-65. DOI: 10.1039/QR9672100043

[49] Anderson KM, Day GM, Paterson MJ, Byrne P, Clarke N, Steed JW. Structure calculation of an elastic hydrogel from sonication of rigid small molecule components. Angewandte Chemie, International Edition. 2008;**47**:1058-1062. DOI: 10.1002/anie.200703785

[50] Saha A, Roy B, Garai A, Nandi AK. Two-component thermoreversible hydrogels of melamine and gallic acid. Langmuir. 2009;**25**:8457-8461. DOI: 10.1021/la900156w

[51] Roy B, Saha A, Esterrani A, Nandi AK. Time sensitive, temperature and pH responsive photoluminescence behaviour of a melamine containing bicomponent hydrogel. Soft Matter. 2010;**6**:3337-3345. DOI: 10.1039/C0SM00036A

[52] Saha A, Roy B, Esterrani A, Nandi AK. Effect of complementary small molecules on the properties of bicomponent hydrogel of riboflavin. Organic & Biomolecular Chemistry. 2011;**9**:770-776. DOI: 10.1039/C0OB00670J

[53] Oster G, Yang NL. Photopolymerization of vinyl monomers. Chemical Reviews. 1968;**68**:125-151. DOI: 10.1021/cr60252a001

[54] Encinas MV, Rufs AM, Bertolotti S, Previtali CM. Free radical polymerization photoinitiated by riboflavin/amines. Effect of the amine structure. Macromolecules. 2001;**34**:2845-2847. DOI: 10.1021/ma001649r

[55] Batchelor RR, Kwandou G, Spicer PT, Stenzel MH. (−)-Riboflavin (vitamin B2) and flavin mononucleotide as visible light photoinitiators in the thiol–ene polymerisation of PEG-based hydrogels. Polymer Chemistry. 2017;**8**:980-984. DOI: 10.1039/c6py02034h

[56] Patra P, Seesala VS, Soni SR, Roy RK, Dhara S, Ghosh A, et al. Biopolymeric pH-responsive fluorescent gel for in-vitro and in-vivo colon specific delivery of metronidazole and ciprofloxacin. European Polymer Journal. 2019;**114**:255-264. DOI: 10.1016/j.eurpolymj.2019.02.037

[57] Wang B, Liu MZ, Liang R, Ding SL, Chen ZB, Chen SL, et al. MMTCA recognition by molecular imprinting in interpenetrating polymer network hydrogels based on poly(acrylic acid) and poly(vinyl alcohol). Macromolecular Bioscience. 2008;**8**:417-425. DOI: 10.1002/mabi.200700176

[58] Hu J, Hou Y, Park H, Choi B, Hou S, Chung A, et al. Visible light crosslinkable chitosan hydrogels for tissue engineering. Acta Biomaterialia. 2012;**8**:1730-1738. DOI: 10.1016/j.actbio.2012.01.029

[59] Choi B, Kim S, Lin B, Wu BM, Lee M. Cartilaginous extracellular matrix-modified chitosan hydrogels for cartilage tissue engineering. ACS Applied Materials & Interfaces. 2014;**6**:20110-20121. DOI: 10.1021/am505723k

[60] Donnelly PE, Chen T, Finch A, Brial C, Maher SA, Torzilli PA. Photocrosslinked tyramine-substituted hyaluronate hydrogels with tunable mechanical properties improve immediate tissue-hydrogel interfacial strength in articular cartilage. Journal of Biomaterials Science. Polymer Edition. 2017;**28**:582-600. DOI: 10.1080/09205063.2017.1289035

[61] Fawzy AS, Nitisusanta LI, Iqbal K, Daood U, Beng LT, Neo J. Chitosan/riboflavin-modified demineralized dentin as a potential substrate for bonding. Journal of the Mechanical Behavior of Biomedical Materials. 2013;**17**:278-289. DOI: 10.1016/j.jmbbm.2012.09.008

[62] Lee HJ, Fernandes-Cunha GM, Myung D. In situ-forming hyaluronic acid hydrogel through visible light-induced thiolene reaction. Reactive and Functional Polymers. 2018;**131**:29-35. DOI: 10.1016/j.reactfunctpolym.2018.06.010

[63] Kim SH, Chu CC. Fabrication of a biodegradable polysaccharide hydrogel with riboflavin, vitamin B2, as a photo-initiator and L-arginine as coinitiator upon UV irradiation. Journal of Biomedical Materials Research Part B: Applied Biomaterials. 2009;**91B**:390-400. DOI: 10.1002/jbm.b.31414

Chapter 3

Biocatalysis and Strategies for Enzyme Improvement

Yauheniya Osbon and Manish Kumar

Abstract

Biotransformation with the help of enzymes can greatly improve the rate and stereospecificity of reactions in organic chemistry. However, the use of organic solvents and harsh conditions in biotechnological applications often correlates with enzyme deactivation or a dramatic drop in catalytic activity. Detailed molecular understanding of the protein structure and conformational dynamics allows us to address such limitations and to finely tune catalytic activity by modifying the solvent, the support, or the active site of the enzyme. Along with physico-chemical methods of enzyme stabilization, such as additive approach, chemical modification, and immobilization of enzymes, approaches of enzyme engineering based on DNA recombination can be used to enhance the performance of biocatalysts. Since successful synthetic and industrial applications of biocatalysts require systems that are not only stable and active, but can also be reused in a continuous flow process reducing the production cost, the goal of this chapter is to introduce the reader to the vast scope of techniques available for enzyme improvement, highlighting their opportunities and limitations for the real-world technological processes.

Keywords: biocatalysis, enzymes, additive approach, PEGylation, enzyme immobilization, adsorption, entrapment, encapsulation, cross-linking, CLECs, CLEAs, nano-biocatalysis, enzyme engineering, rational redesign, site directed mutagenesis directed evolution, DNA shuffling, error prone PCR, saturation mutagenesis, CASTing, *de novo* enzyme design

1. Introduction

Enzymes are biological catalysts that are believed to be the cornerstones of life. They assure metabolic needs of cells and assist in a great range of life-sustaining biochemical reactions. The majority of natural enzymes are highly efficient and can increase the rate of biotransformation up to 10^{17} fold [1]. Enzymes can carry their functions at ambient temperatures and pressures, with a minimum of by-products and waste, leading to the specific product of interest in a single catalyzed step, whereas synthesis of the same product with the means of organic chemistry may require many steps and produce a mixture of undesired isomeric, epimeric, or rearranged compounds [2]. The field of biotechnology strives to exploit isolated enzymes and whole cell cultures as biocatalysts capable of accelerating and refining complex chemical transformations of organic compounds for industrial and synthetic use [3]. Well known examples of such biocatalysts include microbial lipases that are used to synthesize cost-effective biopolymers, biodiesel, pharmaceuticals, and agrochemicals from renewable natural sources, *b*-glycosidases employed in industrial plant biomass saccharification [4, 5] and fungal oxidoreductases that have a potential to become biocatalysts in a bio-based (circular) economy by converting biomass into renewable building blocks for manufacturing biodegradable materials [6]. Unfortunately, the scope of natural enzymes is limited, and certain challenges have to be overcome before we can rely on biocatalysts for efficient, low-cost industrial transformations and greener synthetic chemistry. Such challenges include instability of enzymes *in vitro* (denaturation in high temperatures or extreme pH), low selectivity, product and substrate inhibition, and low reaction yield in non-aqueous solvents [2, 7]. Four general approaches exist to address the above mentioned limitations: additive approach, chemical modification, enzyme immobilization, and protein engineering [8]. While protein engineering is concerned with modifying functional properties of the enzyme at the genetic level, the other three approaches are focused on physico-chemical alterations of the media, enzymatic surface residues, or support material for biocatalyst stability [9].

Using **Figure 1** as a guide, we will review both physico-chemical and functional modification strategies for enzyme improvement, starting with the earliest methods to address solvent-dependent limitations and leading to the most recent technologies, like *de novo* and computational enzyme design [10, 11] (**Figure 1**).

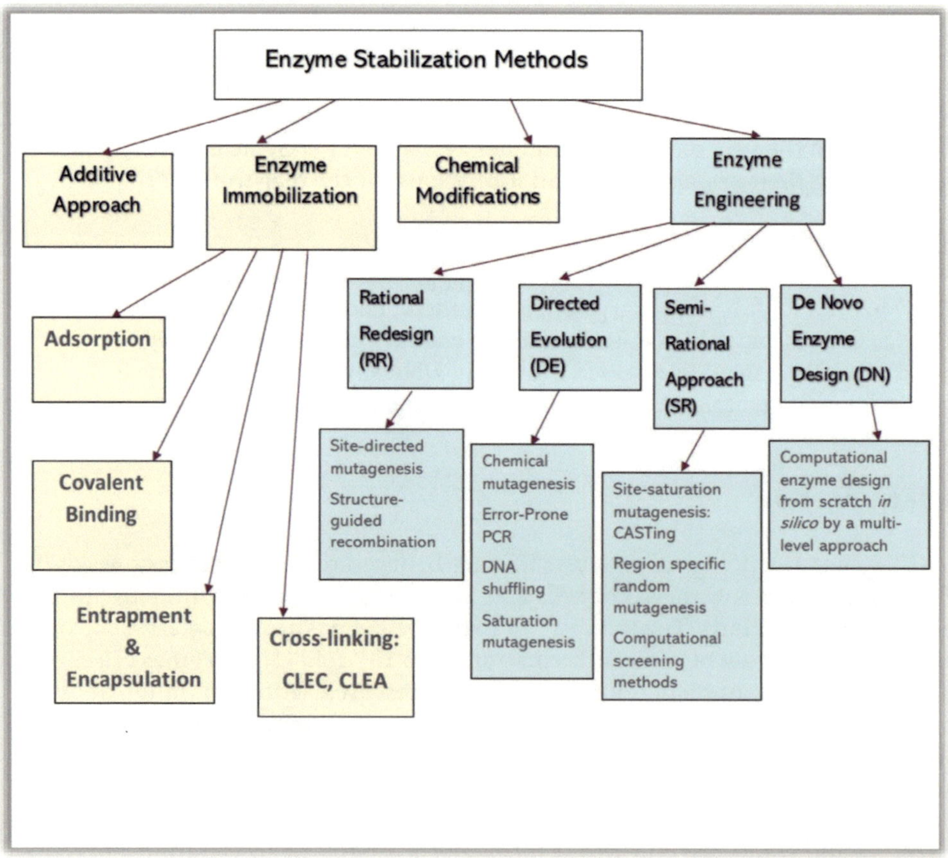

Figure 1.
Enzyme stabilization methods available for improvement of physico-chemical (yellow) and functional (blue) properties of biocatalysts.

2. Improving biocatalysis in organic solvents by additive approach

Enzymes may be remarkable catalysts in biological systems where water is ubiquitous, but they are less suited for biotechnology where organic solvents are largely employed for chemical transformations. For example, the activity of intestinal proteases subtilisin and a-chymotrypsin is reduced 10^4–10^5 times if the enzymes are transferred from aqueous to anhydrous octane media [12]. Knowing that water is essential for structural integrity of many biomolecules and seeing vast experimental evidence of decreased catalytic activity in non-aqueous solvents, scientists have been skeptical about using enzymes as industrial biocatalysts or about using water as a solvent for industrial applications due to undesired hydrolytic side-reactions [3]. However, these challenges were proved to be surmountable when new, improved properties of the enzymes in organic and ionic solvents were discovered several decades ago. In many cases, enzymes that had been stripped of their folded structure in a non-aqueous solution not only became more thermostable and easier to store (due to higher melting points), but also became capable of catalyzing new reactions, impossible in aqueous media. For instance, hydrolases, such as subtilisin, routinely accelerate hydrolysis in aqueous conditions, but in anhydrous solvent, they are capable of catalyzing transesterification reactions [12]. The possibility of using novel, industrially favored substrates as well as the possibility of controlling enzymatic activity and selectivity by finely tuned modifications of the solvent lead to the discovery of numerous approaches to stabilize the enzyme in a non-aqueous solvent [3]. Early trial and error experiments with lyophilized (freeze dried) enzyme powders and solvent additives resulted in the development of empirical strategies like addition of water or water-mimicking solvents and addition of salts for stabilizing biocatalysts. The additive approach currently employs the addition of wide variety of lyophilized chemical substances, also known as lyoprotectants, to the media and still appeals to scientists, as it incorporates the simplicity of use and high efficiency [13, 14].

2.1 Addition of water

The lubricating effects of water on biocatalyst flexibility in organic systems were highlighted in multiple studies [12, 15–19]. For instance, in chymotrypsin activity trials, the amount of residual water retained on the enzyme after the addition of organic component correlated with the catalytic activity of the enzyme [12]. It was later determined that the addition of trace amounts of water, even if the enzyme have been unfolded in organic media, can remediate some of the activity loss: in the experiments with subtilisin Carlsberg suspended in organic solvents, increase in water content from 0 to 1% resulted in the increase of reaction rate 11-fold in isooctane and 50-fold in THF [18]. Moreover, hydration of organic solvent does not prevent the enzyme from acquiring novel properties valuable for synthetic and industrial applications. For example, adding 1% water to glycerol helped to retain the secondary structure of a-chymotrypsin similar to that in aqueous solvent, however, the enzyme stability at high pH was still much greater in 99% organic solvent over that in water [19]. More importantly, while the enzyme suspended in water was fully denatured after 1 min at 100°C, a-chymotrypsin in 99% glycerol retained 80% of its catalytic activity after incubation at 100°C for 10 h [19].

In recent years, water addition strategy has benefitted many promiscuous biocatalyzed synthesis reactions, such as Henry reaction, Michael addition, Mannich reaction, asymmetric aldol reactions, and others [14]. You can refer to the excellent review by Liang and Lin for the empirical data on yield increase in these reactions due to hydration.

It is important to note that while too few water molecules may be not enough to activate biocatalyst in organic solvent, too many water molecules may result in reduced substrate solubility or hydrolytic reactions side product [14].

2.2 Addition of water mimics and lyoprotectants

Just like water that is thought to lubricate the enzyme enhancing protein flexibility with its multiple hydrogen bonds, water-mimicking substances, such as glycerol, formamide, ethylene glycol and formic acid can provide similar hydrogen bonding, while avoiding unintended hydrolysis product [14, 20]. One of the early water mimic studies concluded that adding 0.1% ethylene glycol to the solvents with optimal water content of 0.2% can increase the activity and stereo-selectivity of *Candida cylindracea* lipase [21].

Since enzymes often have to be freeze-dried, it is important to ensure their stability during the long-term storage or temperature changes associated with thawing. The most common lyoprotectant up to date is trehalose sugar that helps to preserve enzyme structure and allows for industrial storage of biocatalysts [18].

Most recent reviews also list organic bases, crown ethers, surfactants, and salts as possible additives used to improve catalytic activity of enzymes in chemical synthesis [14]. For the purpose of this section, we will only cover the addition of salts and the role of ionic interactions in biocatalyst enhancement.

2.3 Addition of salt

In 1994, Khmelnitsky et al. discovered that lyophilization of an enzyme in a salty matrix prior to its suspension in organic media lead to a dramatic enhancement in the rate of catalyzed reaction [22]. In this study, 3750-fold increase in activity of subtilisin Carlsberg was documented when 98% w/w KCl-containing lyophilized enzyme powder was added to hexane, as opposed to salt-less enzyme preparation [22]. The authors explained this phenomenon by the protective ability of the salt that was able to prevent direct contact between enzyme molecules and the organic solvent; however, more recent findings using electron spin resonance spectroscopy suggest that salt-induced ionization stabilizes the charged transition state and thus, increases the polarity of the active site [23]. It is also known that while adding certain ions to enzyme preparation or sometimes directly into the solvent can improve both the reaction rate and enantioselectivity, other ions improve only the rate of the reaction or have no effect on the catalysis [14]. Empirical evidence suggests that only kosmotropic (increasing viscosity of water) salts can stabilize catalysts due to preferential hydration effect that addition of Ca ions is more activating than the addition of Ba, Sr, or Mg divalent metal salts [24, 25], and that by using aqueous solutions of smaller alkali metals or alkaline earth metals rather than hydrating the enzyme with water alone we can markedly increase enantioselectivity of the reaction [26].

Even though additive approach for biocatalysis improvement has offered many successful results, several disadvantages limit its use and call for exploring other methods of enzyme stabilization. Such disadvantages include the fact that the effect of molecular additives varies widely from case to case, depending on the enzymes used, desired substrates and reactions. Since the majority of successful additive methods were discovered by accident, there are no general protocols developed for this approach, and only few stabilizing additives are researched enough for us to clearly understand the molecular mechanisms behind their role in catalysis [14].

3. Chemical modifications to stabilize enzymes

Chemical modification of enzymes is a very common stabilization strategy. In fact, covalent modification by the cross-linking with glutaraldehyde reagent can stabilize almost any enzyme, protecting it from denaturing and other effects of the new solvent [27]. This finding led to the development of carrier-free enzyme immobilization methods (to be discussed in detail in Section 4).

Another popular chemical method involves covalent conjugation with an amphiphilic polymer polyethylene glycol (PEG) and is often referred to as enzyme PEGylation [28]. PEGylating permits binding of specific polymeric functional groups to the free amino groups on the enzyme, creating PEGylated biocatalyst soluble in organic solvents [19]. This method is especially useful for preparation of biopharmaceuticals with high stability and low antigenicity [29]. Additionally, reagent methoxymethyl-PEG (mPEG) bound to the enzyme horseradish peroxidase (HRP) can protect the protein from pH extremes and high temperatures, making HRP particularly useful in industrial and clinical biosensors [30].

Chemical alterations of the enzyme can introduce a new functional group for a covalent attachment or modify one of the reactive side chains. For instance, the treatment of *Candida rugosa* lipase with diethyl *p*-nitrophenyl phosphate modified one of the two reactive functionalities and resulted in a more selective lipase catalyzing a single reaction [31].

Chemical modification makes it possible to introduce and attach a new cofactor, which can in turn induce novel enzyme functions [32]. Lastly, some post-translational enzyme modifications either *in vivo* or *in vitro* have been linked to an enhanced stability of enzymes [33]. For example, when DNA ligase from *Thermus scotoductus* was chemically adenylated, new irreversible covalent binding of the cofactor resulted in structural changes within the active site and overall protein compaction. As the result of this cofactor-induced conformational change, the enzyme gained increased resistance to thermoinactivation [34, 35].

As strategies for enzyme improvement continue to evolve, chemical modification has been rediscovered to become a robust complimentary approach to both protein engineering and immobilization [36].

4. Enzyme immobilization

It is clear that while additive or chemical approaches represent a simple and attractive route for a small-scale chemical synthesis, they have to be supplemented or substituted with other enzyme-stabilizing approaches when applied to complex enzyme systems or whole-cell catalysts in biotechnology.

Much like in a living organism, where enzymes are associated with a membrane or a cell structure that ensures their stability, in industrial setting, it is often necessary to anchor the enzyme to a certain area of a reactor in order to stabilize and reuse the same catalytic device over and over [8]. Enzyme immobilization approach usually achieves this goal by constraining enzymes to a more stable support (a carrier), thus, creating insoluble heterogeneous catalyst of native conformation with reduced flexibility [37]. Not only immobilized catalyst is less likely to be deactivated in organic solvent, is not perturbed by lyophilization [38], is more resistant to sheer stress and high temperatures [39, 40], but it is also less costly, as it is recycled in a continuous fixed-bed process and allows for an efficient enzyme purification and recovery with the help of selective adsorbents [41, 42].

Biocompatible carrier can be chosen from almost any organic polymer or inorganic material that is inert [43]. The requirements to the ideal carrier also include: affordability (the cost of the enzyme and the carrier should not be more than a few percent of the total production cost), stability, physical strength, regeneration ability, as well as the ability to aid in catalytic functions [38, 44], promoting enzyme specificity and reducing product inhibition and non-specific interactions [45]. Moreover, the carrier has to match certain surface properties of the enzyme (e.g. polar groups of amino acids or apolar surface areas), have high superficial density of reactive groups [46], and have a large surface area (e.g. contain high number of small-sized particles or large dimensional pores) [47]. In some cases, where stronger covalent binding with the enzyme is desired, reactive functional groups have to be chemically introduced as monomers in a polymeric carrier matrix [48]. All the above requirements for the carrier are not easy to meet, and sometimes carrier-less approach is used, where enzyme molecules are cross-linked to each other, providing support [38]. Fundamental strategies of biocatalyst immobilization include adsorption, covalent binding, cross-linking, and entrapment/encapsulation [8, 47, 49] (**Figure 2**). In this section, we will review these approaches and their successful combinations, as well as some promising "smart" enzyme immobilization techniques with possible applications.

4.1 Adsorption

More than a century ago, it was empirically proven that invertase enzyme can be adsorbed onto a solid support like charcoal or aluminum hydroxide with its catalytic activity remaining similar to that in its free soluble state [50]. Since then, the methods of adsorption of enzymes onto carriers may have evolved, but the principle has remained the same: intermolecular forces between the enzyme and support result in protein accumulation on a solid surface [8]. Adsorption forces are rather mild and generally involve non-specific van der Waals, hydrophobic, hydrogen bond (H-bond) and ionic interactions [42]; however, high coverage and stronger adherence can be achieved by choosing the right carrier based on enzyme's surface

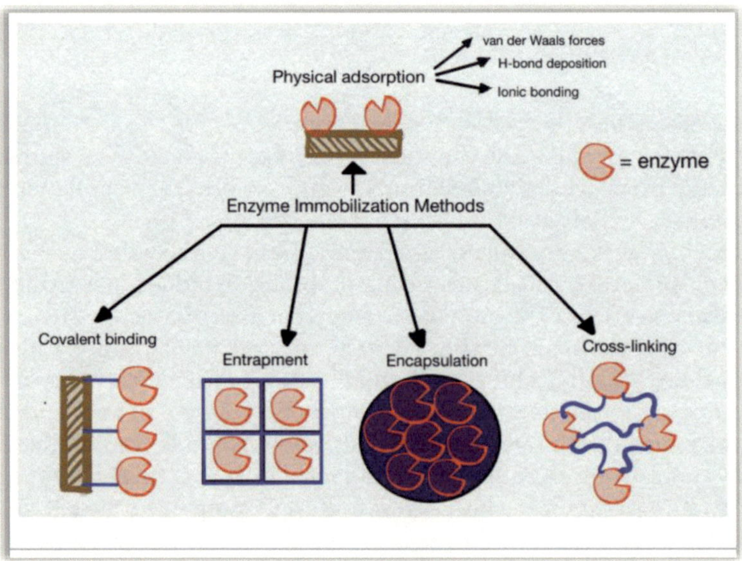

Figure 2.
Fundamental methods of enzyme immobilization.

charge and polarity [51]. For instance, a hydrophobic carrier will work well with the enzyme of large lipophilic surface area due to entropy and hydrophobic effect [47]. Alternatively, H-bonding to hydrophilic carriers (e.g. cellulose, porous glass, silica gel, Avicel, or Celite) is used if the enzyme has hydrophilic surface residues, especially, if the enzyme is glycosylated [47].

The main advantage of adsorption method is its simplicity and low cost [52, 53]. The enzyme can be simply added to the active adsorbent's surface [43], or deposited onto the carrier via evaporation of aqueous phase [54], there is no need to previously modify the ligand [55], or to wash off the enzyme that has not been adsorbed [43]. Typically, mild, reversible interactions of physical adsorption are preferred for immobilization, since they do not change the native structure of the catalyst and do not disturb its active enzymatic sites [53]. The disadvantages of adsorption include the fact that it only proceeds in organic solvents due to intrinsic solubility of enzymes in such media; otherwise, enzyme leaching is unavoidable either as a result of H-bonds with water or due to ion exchange for the enzymes immobilized by ionic forces [47]. Apart from enzyme leaching as a function of time [8], in a physical process such as adsorption, active site blockage due to nonspecific interaction with the carrier can greatly reduce the activity of immobilized catalyst, especially if the substrate is large [49, 56].

In order to improve the stability and activity outcomes of non-specific adsorption, multiple specific adsorption methods were proposed in recent years, including biospecific adsorption that involves immobilized antibodies, affinity adsorption that uses carrier-bound immobilized dyes as ligands, coordination adsorption where immobilized transition metal ions interact with amino acid residues on the enzyme, and others [57].

4.2 Covalent binding

It was discovered in 1960s that certain pre-functionalized carriers covalently bound to the enzyme (native or modified) can act as a scaffold for enzyme stabilization, while substantially improving its performance as a biocatalyst [42]. Covalent forces between the functional groups of the support matrix (whether it is an inner wall of a bioreactor, or a packed-bed industrial reactor filled with glass or biopolymeric beads [8] and reactive active amino acid residues on the enzyme surface can create extremely strong linkages [58]. For example, pre-treatment of glucose oxidase from *Aspergillus niger* with periodic acid helped to activate its carbohydrate residues for new covalent linkages to a hydrophobic polymer *p*-aminostyrene. As a result, immobilized glucose oxidase retained its full activity and even gained higher thermostability at 60°C compared to that of a native soluble enzyme [59]. Unlike adsorption, covalent binding can be performed in any solvent; however, it is a method of choice for immobilizing enzymes in aqueous solvents or under denaturing conditions [47]. The rigid covalent binding prevents enzyme leaching caused by non-specific interactions with water and locks the enzyme in its native conformation, resisting thermoinactivation [60]. Similar to physical adsorption, both hydrophilic and hydrophobic carriers can be used for immobilization [47]. Carrier activation may also require long or short spacer molecules between the carrier and the enzyme, in order to provide more accessible catalytic sites and to attach desired reactive groups to the matrix [55]. Most commonly introduced reactive groups include aldehydes or epoxides to be attacked by nucleophilic amino groups on the protein, which is followed by instant reduction of the product and irreversible attachment of the enzyme to the carrier [42, 48]. Nevertheless, such chemical modifications can be harsh for the three-dimensional protein structure, especially, if chemical microenvironment of the enzyme (including possible storage additives) is interfering. The main disadvantages of immobilization by covalent binding include non-uniform attachment if the bond density is too low [61], or, if the bond density is too high, the risk of immobilized catalyst being irreversibly deactivated and rendered unusable [37].

4.3 Entrapment and encapsulation

Entrapment and encapsulation within the polymeric matrix are immobilization techniques that, similar to physical adsorption, employ non-covalent interactions [49]. Unlike in surface adsorption, the support matrix is not pre-fabricated but is synthesized at the same time as the enzyme is being entrapped or encapsulated *in situ* [38].

In the entrapment method, a catalyst (soluble or insoluble) is dispersed in a solution of a monomer or polymer of low molecular weight, and later becomes entrapped in a matrix formed by hydrolytic polymerization [42]. Inside the polymer lattice, almost every side chain on the enzyme surface physically interacts with the support material around it, while allowing small substrate product molecules move in and out of the complex through the pores [43]. As an attempt to modulate the porosity and diffusion pattern, silica sol–gels of varying densities (e.g. xerogels, ambigels, aquagels, or aerogels) with additives to create hydrophilic or hydrophobic surfaces are commonly used [62]. When xerogels of high density entrap one or more enzymes by hydrolytic polymerization, the substrate selectivity of the extremely small pores can be particularly useful for the development of biosensors [63, 64]. Entrapment matrices with hydrophobic surfaces can activate lipases, the most used enzymes in synthetic organic chemistry [65]. Moreover, catalytic activity of *Candida antarctica* lipase entrapped in a hydrophobic sol-gel can be improved 2–8-fold compared to that of non-immobilized freeze-dried lipase powder [66]. Nonetheless, the practical application of entrapment is limited because of the requirements, like small substrate size and delicate balance between physical properties of the matrix and enzymatic activity [43]. Insufficient substrate interaction and leakage of enzymes due to continuous use are major disadvantages of the entrapment method [67].

Alternatively, enzymes can be encapsulated within a semi-permeable polymeric membrane [43]. Immobilization by encapsulation occurs when microcapsule walls are formed around enzymes as a result of polymer desolvation [68]. The capsule can provide desired chemical microenvironment for a specific enzyme in terms of pH, temperature and solvent stability, very similar to the microenvironment in the living cell [69]. Depending on the material used, the membrane can be permanent or non-permanent. Non-permanent membranes can be formed by liquid surfactants, while permanent membranes are often made of polystyrene, cellulose, gliadin, nylon, and other materials used to encapsulate pharmaceuticals, food, cosmetics, and chemicals [68, 70].

Although high enzyme concentrations and large surface areas of encapsulated biocatalyst can ensure faster reaction rates [71], similar to enzyme entrapment, traditional method of microencapsulation can only be applied to a limited number of enzymes with small substrates [62]. In recent years, the structure of microcapsule from a hollow bead evolved to a complex multilevel three-dimensional sphere. Such biomimetic capsule design often allows to adjust permeability in favor of mass transfer of substrate and product, while dramatically improving catalytic activity of encapsulated enzymes [72].

4.4 Carrier-free immobilization by cross-linking

Carrier-free immobilization method was developed as a way to address the issue of inevitable activity drop associated with the carrier attachment. The non-catalytic part often represents 90–99% of the immobilized catalyst [37], and although it provides the catalytic element with enhanced stability and longer half-life, it is also responsible for unwanted mass-transfer effects, low product yields and higher production cost [73].

It was first reported by Yale scientists in 1964 that di-functional reagent glutaraldehyde, normally used as a fixative for electron microscopy slide preparation, can form irreversible covalent linkages between enzyme molecules. When carboxypeptidase-*a* enzyme crystals treated with glutaraldehyde were tested by diffraction techniques, cell dimensions of the enzyme remained very similar to that of its native state, even after mechanical stress and changing the solvent; moreover, the enzyme tested retained most of its catalytic activity [27]. Therefore, pure enzyme in almost any form (solubilized, crystallized, aggregated or atomized) now can be chemically modified to act as its own carrier [74]. The mechanism of this reaction involves the formation of Schiff's bases between carbonyls on the cross-linker and free amino groups on the enzyme's lysine residues together with a pH-dependent Michael 1,4-addition to *a,b*-unsaturated aldehydes [75].

Two most popular carrier-less preparations that can be used in industrial and pharmaceutical synthesis are cross-linked enzyme crystals (CLEC) and aggregates (CLEA) [76]. CLECs, obtained by crystallization of the pure enzyme prior to cross-linking [77], according to the patent of 1997, are able to retain at least 91% of enzyme's initial catalytic activity if incubated with a protease for 3 h, compared to unmodified enzyme that loses at least 94% of the activity under the analogous conditions [78]. Resistance to proteases also allows to administer CLEC biocatalyst orally as a longer-acting drug [79]. As for the industrial uses, advantages of CLECs, like possibility of indefinite storage at room temperature, their near-maximum catalytic activity in harsh conditions, and high tolerance to organic solvents [80], allow for various potential applications, including microporous phase for chromatography and environmental toxicity biosensors [81, 82]. And yet, CLEC method requires highly purified enzyme for crystallization, which translates into higher expense and renders CLEC superseded by closely related technique CLEA [37].

CLEAs, first described in 2000, are prepared by aggregating enzymes with precipitants, like acetone, ethanol, or ammonium sulfate, and then by cross-linking the aggregates with glutaraldehyde or dextran polyaldehyde [38, 83]. After the protein is cross-linked with the reagent, sometimes protective reagents and other additives can be used to create a "tailor-made" biocatalyst [47]. For instance, knowing that surfactants, amines, and crown ethers can activate lipases, one study prepared CLEAs of seven microbial lipases by co-precipitation with different additives. As a result, two of the lipases were hyper-activated (had 2–3 times higher activity than that of a native enzyme) in the presence of SDS surfactant, demonstrating that precipitation with the right additive can lock the catalyst in the preferred conformation [84].

Apart from broad applicability, high stability, and possibility of catalytic hyperactivation by chemical optimization, the CLEA technology eliminates laborious enzyme purification and crystallization steps required for CLECs as well as the expense of using a carrier required for other immobilization methods [85]. This makes CLEA protocols reproducible in almost any lab that has a relatively crude enzyme sample [49]. Interestingly, crude samples containing several enzymes can be used to make combi-CLEAs that recreate multistep, multi-enzyme cascade processes for biotechnological or clinical applications [85].

4.5 Combinatory physicochemical approaches

Rational combinations of immobilization methods described above together with chemical enzyme modifications (see Section 3) often have synergistic effect [33].

Therefore, combinatory approaches have been increasingly applied with the goal of designing an improved, robust catalyst from the native enzyme by physicochemical means at a low cost. For instance, physical adsorption, a simple and less expensive immobilization technique [52], would be more commercially viable if it could avoid desorption of enzyme immobilized by weak, non-specific forces. It was reported that entrapment of *Candida* lipase prior to its co-adsorption with lipase-activating chemical additives not only protected the enzyme from desorption, but also was a cheaper and more efficient method of biodiesel production from waste cooking oil, compared to immobilization methods described in previous studies [57]. Combination of enzyme cross-linking prior to entrapment in sol–gel beads was determined to be a viable technique for stabilizing enzymes in glucose biosensors [67]. In some cases, filtration-enabled reusability of CLEAs may be problematic due to similar size of the particles and the substrate. To ensure easy enzyme separation, it was proposed to encapsulate the enzyme solution in a soft porous membrane and thus, create particles of desired size prior to their aggregation and cross-linking [44].

With a number of successful techniques available and so many variables involved in the trade-off between stability and activity of immobilized enzyme, choosing the most efficient approach for a less-studied enzyme can be a problem. In this case, combination of computational analysis with experimental methods, known as structure-based immobilization, can be of assistance [47]. For instance, GRID computational analysis of the functional groups on the enzyme can help to locate nucleophilic amino groups on the lysine residues that would be involved in the covalent binding. If such groups are present in the close proximity of the active site, non-covalent techniques, like adsorption, entrapment, or encapsulation should be considered. Additionally, hydrophobic and hydrophilic surfaces of the enzyme can be established in order to choose the appropriate support material [86].

According to bibliometric analysis of 2019, new trends in enzyme immobilization research can be seen from the top-50 author-keywords list, based on the works published globally in the last 5 years. The top directive terms included: gold nanoparticles, meso-porous silica, magnetic nanoparticles, response surface methodology, glucose biosensor, and cross-linked enzyme aggregates (CLEAs) [74]. It is evident, that immobilization on nano-particles, nano-fibers, and nano-gels is a field of special interest, due to high adaptability, high retention of activity, and effortless enzyme separation and recycling [49, 87]. The main advantage of nanostructures is their high surface to volume ratio, where decreasing size of the carrier allows progressive exposure of the enzyme to reaction media, making nano-immobilization a method of choice for development of powerful enzyme-based fuel cells [88]. Magnetic nanoparticles (e.g. mCLEA) that can be functionalized for enzyme isolation by magnetic decantation or used in magnetically stabilized reactor beds are part of a new, promising approach, known as "smart" enzyme immobilization [89]. Finally, there is a possible turn of interest to a matrix algebra- and statistics-based response surface method (RSM) that can be used for optimizing operational conditions of immobilized biocatalysts [90].

5. Enzyme engineering

"What I cannot create, I do not understand"
Richard Feynman

Enzyme engineering approach originated as an ultimate challenge to test our understanding of protein structure and function in early 1980s [91]. Since then, the field of biocatalysis have been revolutionized by the advances of recombinant DNA technology, use of computational tools, and modern bioengineering. Alterations of the primary protein structure can be tailored not only to stabilize the enzyme, but also to broaden the substrate range, optimize catalytic performance, and obtain products of high value for various biotechnological applications. Enzyme designers can modify or develop a novel biocatalyst with a new primary function [3] by using one of the following approaches: rational redesign (RR), directed evolution (DE), semi-rational redesign (SR), or *de-novo* enzyme design (DN) [92]. Enzyme engineering methods are laborious and all have the following pre-requisites in common: enzyme encoding gene of interest, microbial or yeast expression system, and sensible screening tools for mutant detection [93]. In this section, we will briefly review these methods and their possible applications in fine chemical synthesis.

5.1 Rational redesign (RR)

RR strategies for enzyme improvement genetically modify the existing biocatalyst based on known structural criteria. To begin with, different pieces of data, like protein structure obtained by X-ray diffraction techniques, molecular models based on computational algorithms, and biochemical specifics, like locations of interacting ligand residues obtained by NMR-analysis, must be evaluated to propose rational genetic alterations [94]. The method of choice for introducing specific alterations is site-directed mutagenesis (SDM) [2]. Most commonly, beneficial mutations are induced with the goal of changing catalytic mechanism, reinforcing the promiscuous reaction, altering substrate/cofactor specificity and improving the overall stability of the biocatalyst [93]. To ensure the specificity of mutations, the majority of RR techniques employ polymerase chain reaction (PCR) with the primers (short sequences of synthetic DNA complementary to the template gene) that have been modified using a known mutant codon or codons [2].

Two common methods of inserting/deleting specific amino acids into/from the target gene by SDM are overlap extension and whole plasmid single round PCR [10]. The former method uses two pairs of primers with one of the primers in each pair containing a modified (mutant) codon, which results in a PCR-produced heteroduplex plasmid with overlapping breaks. The later method is simplified by using a "QuikChange Site-Directed Mutagenesis Kit" patented by Stratagene and adding two primers with desired mutations that are complementary to the opposite strands of the DNA template. The mutant plasmid obtained by PCR is then nicked by a specific restriction enzyme to be repaired upon transformation into competent cells [95]. A more efficient one-step version of QuikChange method suitable for single or multiple-site insertions and deletions was described in [96]. Successful applications of SDM include the induction of acid-resistance and 16.7-fold higher catalytic activity in α-amylase used for industrial sugar and detergent production [97], thermostability enhancement in sucrose isomerase producing a non-cariogenic, nutritional sugar isomaltulose that slows down the rate of insulin release in diabetes management [98], and even attempts to design high efficiency H_2-producing cyanobacterial cells to make biofuels [94]. At the same time, numerous attempts of RR fail due to insufficient knowledge of mechanisms responsible for the specific structure–function relationships. The process of SDM is often too tedious and expensive, requiring mutant enzyme confirmation by sequencing and purification [99].

5.2 Directed evolution (DE)

In contrast to RR, the methods of DE can be applied to the enzymes even in the absence of existing structural and mechanistic data. DE relies on accelerating

Darwinian evolution in laboratory with the means of mutation and random recombination, followed by multiple rounds of selective "molecular breeding" [100]. Three steps common for all DE methods are: (1) construction of mutant library by the means of random mutagenesis or *in-vitro* recombination, (2) screening/selection of mutants with the desired properties via high throughput assays, and (3) improved protein gene isolation [49, 92]. Initial generation of molecular diversity can be achieved by methods like chemical mutagenesis, error-prone PCR (epPCR), gene site saturation mutagenesis, and DNA shuffling [2, 49].

Chemical mutagenesis is used extensively for a "food grade" enzyme improvement: avoiding the introduction of heterologous DNA, it applies chemical agents like ethyl methyl sulfonate or nitrous acid to bacterial strains lacking DNA repair mechanisms, generating random mutations [101].

EpPCR method introduces random changes in a catalyst encoding gene by using error-prone Taq DNA polymerase for the PCR process. Unlike in chemical mutagenesis, the rate of mutation can be controlled by modifying PCR conditions. For instance, it is possible to introduce an average of one amino acid substitution per PCR cycle, and 3–7 cycles are usually enough to improve the thermostability and enantioselectivity of the enzyme [2, 102].

DNA shuffling method relies on in vitro DNA recombination of closely related parental genes either obtained from different organisms or produced by epPCR [92]. The genes are digested with DNaseI or with a mixture of restriction endonucleases to yield random, small fragments that will be purified and reassembled by epPCR, where fragments cross-prime each other [95]. Therefore, genes from multiple parents, including different species, can be shuffled in a single step, sometimes resulting in a hybrid DNA with unique, novel traits not expressed in either parent [103]. Nevertheless, DNA shuffling cannot induce drastic functional changes, as these are known to require considerable evolutionary changes in polypeptide backbone [99].

The technique known as gene site saturation mutagenesis can be used for the replacement of each amino acid of a protein with each of the other 19 amino acids occurring in nature [2]. For instance, PCR amplification with a mixture of 64 different forward and 64 different reverse primers would be necessary to randomize one codon in the enzyme gene (based on 4 letters of genetic alphabet and 3 nucleotides in one codon), but one could eliminate stop codons by restricting the third nucleotide to G or C and use a mixture of 32 forward and reverse degenerate primers. Statistically, the size of the mutant library obtained by this method can be calculated as 20^n, where n is the number of amino acid residues in the protein of interest [2]. Hence, the biggest limitation of DE methods is the requirement of an efficient high throughput screening process for the mutant libraries that even with millions of variants, still sample only a very small fraction of the enzyme sequence [104]. Nevertheless, the field of DE is actively developing, and recent publications report achievements, such as induction of pH-, temperature- and oxidation-tolerance in a catalyst [105, 106], including the development of a novel enzyme structure as a result of a new function acquired by DE [107].

5.3 Semi-rational (SR) and *de novo* (DN) design approaches

Addressing major limitations of both DE and RR, SR design creates smaller, manageable mutant libraries based on the structural/biochemical data or computational predictive algorithms [104]. In other words, if the 3-D structure of an enzyme is available, random mutagenesis can be focused on a specific site (usually within the active site) in the protein sequence, leading to higher probability of identifying the key amino acids, randomly replacing them or reinforcing them via cumulative effects [3].

Site saturation mutagenesis rationally applied to a specific codon or codons is an example of a semi-rational approach. In the method known as CASTing (combinatorial active site saturation test) several amino acids of the active site are targeted and mutated one by one or in combination [108]. Combinatorial saturation of residues can produce mutants improved by synergistic effects. However, these vast multiple-site saturation libraries are almost impossible to screen with traditional DE means. In this case, computational algorithms are used to virtually screen the library, eliminating the mutants that have been misfolded due to unfavorable amino acid interactions [109].

"Region specific random mutagenesis" is another semi-rational method that employs SDM performed with 64 different forward and reverse primers targeting a single codon to be randomized based on its importance for enzyme structure and function [95].

There is no doubt that the fine-tuning of engineered enzymes benefits from the use of combinatorial approaches. SR design has a great potential to create specialized biocatalysts for the industrial use. For example, improved uronate dehydrogenase enzyme of high thermostability can be used to catalyze the production of glycaric acids (top-value chemical precursors for greener and less expensive biofuel synthesis) [110]. Nevertheless, even with reduced mutant libraries, SR approaches are not straight forward. Sometimes, molecular interactions and cascade effects far away from the active site are responsible for certain catalytic effects [3].

In cases when enzyme structures do not appear to be optimized for a specific chemical transformation, synthesizing enzyme of a completely different structure *de novo* may be the most reasonable solution. In 2008, DN engineering of artificial enzyme Kemp eliminase from scratch provided the evidence that it is possible to create a new functional enzyme using a multi-level approach that involves quantum mechanical (QM) calculation, computational algorithms, and directed evolution [111]. Generalized DN workflow starts with the engineering of a minimal ideal active site containing catalytic machinery and interaction residues, optimized by QM into a model structure called "theozyme". The theozyme is then matched to a pre-existing protein scaffold by a hashing algorithm. After the scaffold-theozyme structure is stabilized, the enzyme is ranked *in silico* based on its geometry and binding energy and can be empirically tested. If tested active, but inefficient, the catalytic activity can be improved by several rounds of DE [11]. DN development would not have been possible without computational tools, like METAL SEARCH, DEZYMER, ORBIT, ROSETTA match, ROSETTA design, and MODELER (see [49, 109] for details).

Although it is able to provide predictive frameworks for rational *in silico* engineering and generate more focused, "smart" libraries, computational enzymology is still in its infancy. Combinatorial strategy where directed evolution is integrated with rational optimization remains a method of choice for protein engineering [104].

6. Conclusions

While enzymes have been involved in commercial production processes for centuries, their vast potential for a large scale chemical synthesis and industrial applications was not fully realized until better empirical models and methods of biocatalyst stabilization were developed using a trial and error approach. In this chapter, we reviewed fundamental strategies for enzyme improvement, such as chemical modification, additive approach, enzyme immobilization, and protein engineering. It appears that enzyme immobilization is currently considered to be the most promising strategy for obtaining industrial biocatalysts with controlled, more specific substrate interactions, resistance to denaturation, and high product yield at low cost [74]. At the same time, enzyme engineering methods recently made numerous successful advances to redesign existing enzymes on the level of their primary structure using targeted random mutagenesis, *in vitro* recombination, and various computational tools. Although there is high demand for such specialized, robust biocatalysts, they are generally produced as soluble enzymes, not reusable in the industrial synthesis. Therefore, integration of physico-chemical methods and protein engineering is possibly the most efficient strategy for creating a powerful, recyclable biocatalyst fit for the real-world biotechnological processes.

Conflict of interest

To our knowledge, there has been no conflict of interest.

Notes/thanks/other declarations

We would like to thank our families for the support and encouragement.

Author details

Yauheniya Osbon and Manish Kumar*
Department of Biology, Texas State University, San Marcos, Texas, USA

*Address all correspondence to: m_k135@txstate.edu

References

[1] Benkovic S, Hammes-Schiffer S. A perspective on enzyme catalysis. Science. 2003;**301**(5637):1196-1202. DOI: 10.1126/science.1085515

[2] Glazer AN, Nikaido H. Microbial Biotechnology : Fundamentals of Applied Microbiology [Internet]. Cambridge; New York: Cambridge University Press; 2007. pp. 398-429. DOI: 10.1017/CBO9780511811227.012

[3] Tunon I, Moliner V. Simulating Enzyme Reactivity: Computational Methods in Enzyme Catalysis, RSC Theoretical & Computational Chemistry Series. England: Royal Society of Chemistry; 2017. DOI: 10.1039/9781782626831-FP001

[4] Abir BB, Nadine MSM, Alaa A-A. An organic solvent-stable lipase from a newly isolated *Staphylococcus aureus* ALA1 strain with potential for use as an industrial biocatalyst. Biotechnology and Applied Biochemistry. 2016;**63**(3):378. DOI: 10.1002/bab.1381

[5] Matsuzawa T, Watanabe M, Yaoi K. Improved thermostability of a metagenomic glucose-tolerant beta-glycosidase based on its X-ray crystal structure. Applied Microbiology and Biotechnology. 2017;**101**(23-24):8353-8363. DOI: 10.1007/s00253-017-8525-9

[6] Martínez AT, Ruiz-Dueñas FJ, Camarero S, Serrano A, Linde D, Lund H, et al. Research review paper: Oxidoreductases on their way to industrial biotransformations. Biotechnology Advances. 2017;**35**:815-831. DOI: 10.1016/j.biotechadv.2017.06.003

[7] Kaul P, Asano Y. Strategies for discovery and improvement of enzyme function: State of the art and opportunities. Microbial Biotechnology. 2012;**5**(1):18-33. DOI: 10.1111/j.1751-7915.2011.00280.x

[8] Minteer SD. Enzyme Stabilization and Immobilization: Methods and Protocols. New York, N.Y.: Humana Press, Springer; 2017. DOI: 10.1007/978-1-4939-6499-4

[9] O'Fagain C. Enzyme stabilization—Recent experimental progress. Enzyme and Microbial Technology. 2003;**33**(2-3):137-149. DOI: 10.1016/S0141-0229(03)00160-1

[10] Turanli-Yildiz B, Alkim CZ, Cakar P. In: Kaumaya P, editor. Protein Engineering Methods and Applications, Protein Engineering. Rijeka: InTech; 2002. ISBN: 978-953-51-0037-9

[11] Barrozo A, Borstnar R, Marloie G, Lynn Kamerlin SC. Computational protein engineering: Bridging the gap between rational design and laboratory evolution. International Journal of Molecular Sciences [Internet]. 2012;**13**(10):12428-12460. DOI: 10.3390/ijms131012428

[12] Zaks A, Klibanov AM. The effect of water on enzyme action in organic media. Journal of Biological Chemistry. 1988;**263**:8017-8021. DOI: 10.1073/pnas.82.10.3192

[13] Klibanov A. Why are enzymes less active in organic solvents than in water? Trends in Biotechnology. 1997;**15**(3):97-101. DOI: 10.1016/S0167-7799(97)01013-5

[14] Liang Y-R, Wu Q, Lin X-F. Effect of additives on the selectivity and reactivity of enzymes. Chemical Record. 2017;**17**(1):90-121. DOI: 10.1002/tcr.201600016

[15] Lee CS, Ru MT, Haake M, Dordick JS, Reimer JA, Clark DS. Multinuclear NMR study of enzyme hydration in an organic solvent. Biotechnology and Bioengineering. 1998;**57**(6):686. DOI: 10.1002/(SICI)

1097-0290(19980320)57:6<686::AID-BIT6>3.0.CO;2-H

[16] Kamal J, Xia T, Pal S, Zhao L, Zewail A. Enzyme functionality and solvation of Subtilisin Carlsberg: From hours to femtoseconds. Chemical Physics Letters. 2004;**387**(4-6):209-215. DOI: 10.1016/j.cplett.2004.02.011

[17] Hudson EP, Eppler RK, Clark DS. Biocatalysis in semi-aqueous and nearly anhydrous conditions. Current Opinion in Biotechnology. 2005;**16**(6):637-643. DOI: 10.1016/j.copbio.2005.10.004

[18] Hudson EP, Eppler RK, Beaudoin JM. Active-site motions and polarity enhance catalytic turnover of hydrated subtilisin dissolved in organic solvents. Journal of the American Chemical Society. 2009;**131**(12): 4294-4300. DOI: 10.1021/ja806996q

[19] Castro GR, Knubovets T. Homogeneous biocatalysis in organic solvents and water-organic mixtures. Critical Reviews in Biotechnology. 2003;**23**(3):195-231. DOI: 10.1080/714037689

[20] Rupley JA, Careri G. Protein hydration and function. Advances in Protein Chemistry. 1991;**41**:37-172. DOI: 10.1016/S0065-3233(08)60197-7

[21] Gubicza L, Kelemenhorvath I. Effect of water-mimicking additives on the synthetic activity and enantioselectivity in organic solvents of a Candida Cylindracea lipase. Journal of Molecular Catalysis. 1993;**84**(1):L27-L32. DOI: 10.1016/0304-5102(93)80080-E

[22] Khmelnitsky YL, Welch SH, Dordick JS, Clark DS. Salts dramatically enhance activity of enzymes suspended in organic solvents. Journal of the American Chemical Society. 1994;**116**(6): 2647-2648. DOI: 10.1021/ja00085a066

[23] Ru M, Dordick J, Reimer J, Clark D. Optimizing the salt-induced activation of enzymes in organic solvents: Effects of lyophilization time and water content. Biotechnology and Bioengineering. 1999;**63**(2): 233-241. DOI: 10.1002/(SICI)1097-0290(19990420)63:2<233::AID-BIT12>3.0.CO;2-S

[24] Ru M, Hirokane S, Lo A, Dordick J, Reimer J, Clark D. On the salt-induced activation of lyophilized enzymes in organic solvents: Effect of salt kosmotropicity on enzyme activity. Journal of the American Chemical Society. 2000;**122**(8):1565-1571. DOI: 10.1021/ja9935198

[25] Sasaki T, Kise H. Increase of catalytic activity of α-chymotrypsin by metal salts for transesterification of an amino acid ester in ethanol. Bioscience, Biotechnology & Biochemistry. 1997;**61**:1196-1197. DOI: 10.1271/bbb.61.1196

[26] Okamoto T, Yasuhito E, Ueji S. Metal ions dramatically enhance the enantioselectivity for lipase-catalysed reactions in organic solvents. Organic & Biomolecular Chemistry. 2006;**4**(6):1147-1153. DOI: 10.1039/b514885e

[27] Quiocho FA, Richards FM. Intermolecular cross linking of a protein in the crystalline state: Carboxypeptidase-A. Proceedings of the National Academy of Sciences of the United States of America. 1964;**52**(3):833. DOI: 10.1073/pnas.52.3.833

[28] Jeng F-Y, Lin S-C. Characterization and application of PEGylated horseradish peroxidase for the synthesis of poly(2-naphthol). Process Biochemistry. 2006;**41**(7):1566-1573. DOI: 10.1016/j.procbio.2006.02.021

[29] Lundblad RL, Bradshaw RA. Applications of site-specific chemical modification in the manufacture of biopharmaceuticals: I. An

overview. Biotechnology and Applied Biochemistry. 1997;**26**(3):143-151. DOI: 10.1111/j.1470-8744.1997.tb01322.x

[30] Garcia D, Ortéga F, Marty J-L. Kinetics of thermal inactivation of horseradish peroxidase: Stabilizing effect of methoxypoly(ethylene glycol). Biotechnology and Applied Biochemistry. 1998;**27**(1):49-54. AN:edsagr.US201302886708

[31] Fadnavis NW, Babu RL, Vadivel SK, Deshpande AA, Bhalerao UT. Lipase catalyzed regio- and stereospecific hydrolysis: Chemoenzymatic synthesis of both (R)- and (S)-enantiomers of α-lipoic acid. Tetrahedron: Asymmetry. 1998;**9**(23):4109-4112. DOI: 10.1016/S0957-4166(98)00447-9

[32] Sinha RR, Imperiali B. Pyridoxamine-amino acid chimeras in semisynthetic aminotransferase mimics. Protein Engineering. 1997;**10**(6): 691-698. DOI: 10.1093/protein/10.6.691

[33] Cowan DA, Fernandez-Lafuente R. Enhancing the functional properties of thermophilic enzymes by chemical modification and immobilization. Enzyme and Microbial Technology. 2011;**49**(4):326-346. DOI: 10.1016/j.enzmictec.2011.06.023

[34] Georlette D, Blaise V, Dohmen C, Gerday C, Feller G, Bouillenne F, et al. Cofactor binding modulates the conformational stabilities and unfolding patterns of NAD+-dependent DNA ligases from *Escherichia coli* and *Thermus scotoductus*. Journal of Biological Chemistry. 2003;**278**(50):49945-49953. DOI: 10.1074/jbc.M307761200

[35] Georlette D, Blaise V, Gerday C, Feller G, Bouillenne F, Damien B, et al. Adenylation-dependent conformation and unfolding pathways of the NAD +-dependent DNA ligase from the thermophile *Thermus scotoductus*. Biophysical Journal. 2004;**86**(2):1089-1104. DOI: 10.1016/S0006-3495(04)74184-3

[36] DeSantis G, Jones JB. Chemical modification of enzymes for enhanced functionality. Current Opinion in Biotechnology. 1999;**10**(4):324-330. DOI: 10.1016/S0958-1669(99)80059-7

[37] Sheldon RA, van Pelt S. Enzyme immobilisation in biocatalysis: Why, what and how. Chemical Society Reviews. 2013;**42**:6223-6235. DOI: 10.1039/c3cs60075k

[38] Sheldon RA. Enzyme immobilization: The quest for optimum performance. Advanced Synthesis & Catalysis. 2007;**349**(8-9):1289-1307. DOI: 10.1002/adsc.200700082

[39] Guisan JM. Immobilization of Enzymes and Cells [Internet], Methods in Biotechnology. Vol. 22. Totowa, NJ: Humana Press; 2006. DOI: 10.1007/978-1-59745-053-9

[40] Cabral JMS, Kennedy JF. In: Gupta MN, editor. Thermostability of Enzymes. Berlin; New York : Springer Verlag; New Delhi: Narosa Pub; 1993. pp. 163-179

[41] Massolini G, Calleri E. Immobilized trypsin systems coupled on-line to separation methods: Recent developments and analytical applications. Journal of Separation Science. 2005;**28**(1):7-21. DOI: 10.1002/jssc.200401941

[42] Cao L. Carrier-bound Immobilized Enzymes: Principles, Application and Design. Weinheim, Germany: Wiley-VCH Verlag GmbH & Co. KGa; 2005. DOI: 10.1002/3527607668

[43] Dwevedi A. Enzyme Immobilization: Advances in Industry, Agriculture, Medicine, and the Environment. Switzerland: Springer; 2016. DOI: 10.1007/978-3-319-41418-8_1

[44] Cao L. Immobilised enzymes: Science or art? Current Opinion in Chemical Biology. 2005;**9**(2):217-226. DOI: 10.1016/j.cbpa.2005.02.014

[45] Datta S, Christena LR, Rajaram YRS. Enzyme immobilization: An overview on techniques and support materials. 3 BIOTECH. 2013;**3**(1):1-9. DOI: 10.1007/s13205-012-0071-7

[46] Mateo C, Palomo JM, Fernandez-Lorente G, Guisan JM, Fernandez-Lafuente R. Review: Improvement of enzyme activity, stability and selectivity via immobilization techniques. Enzyme and Microbial Technology. 2007;**40**:1451-1463. DOI: 10.1016/j.enzmictec.2007.01.018

[47] Hanefeld U, Gardossi L, Magner E. Understanding enzyme immobilisation. Chemical Society Reviews. 2008;**38**(2):453-468. DOI: 10.1039/b711564b

[48] Basso A, De Martin L, Ebert C, Gardossi L, Linda P, Sibilla F. Organically modified xerogels as novel tailor-made supports for covalent immobilisation of enzymes (penicillin G acylase). Tetrahedron Letters. 2003;**44**:5889-5891. DOI: 10.1016/S0040-4039(03)01372-8

[49] Madhavan A, Sindhu R, Binod P, Sukumaran RK, Pandey A. Strategies for design of improved biocatalysts for industrial applications. Bioresource Technology. 2017;**245**:1304-1313. DOI: 10.1016/j.biortech.2017.05.031

[50] Nelson JM, Griffin EG. Adsorption of invertase. Journal of the American Chemical Society. 1916;**38**:1109-1115. DOI: 10.1021/ja02262a018

[51] Cooney MJ, Svoboda V, Lau C, Martin G, Minteer SD. Enzyme catalysed biofuel cells. Energy & Environmental Science. 2008;**1**(3):320-337. DOI: 10.1039/b809009b

[52] Sun J, Jiang Y, Zhou L, Gao J. Research paper: Immobilization of *Candida antarctica* lipase B by adsorption in organic medium. New Biotechnology. 2010;**27**:53-58. DOI: 10.1016/j.nbt.2009.12.001

[53] Jesionowski T, Zdarta J, Krajewska B. Enzyme immobilization by adsorption: A review. Adsorption-Journal of the International Adsorption Society. 2014;**20**(5-6):801-821. DOI: 10.1007/s10450-014-9623-y

[54] Ferrer M, Plou FJ, Fuentes G, Cruces MA, Andersen L, Kirk O, et al. Effect of the immobilization method of lipase from *Thermomyces lanuginosus* on sucrose acylation. Biocatalysis and Biotransformation. 2002;**20**(1):63. DOI: 10.1080/10242420210153

[55] Kirschning A. Immobilized Catalysts : Solid Phases, Immobilization, and Applications. Berlin; New York: Springer; 2004. DOI: 10.1007/b94543

[56] Hernandez K, Fernandez-Lafuente R. Control of protein immobilization: Coupling immobilization and site-directed mutagenesis to improve biocatalyst or biosensor performance. Enzyme and Microbial Technology. 2011;**48**(2):107-122. DOI: 10.1016/j.enzmictec.2010.10.003

[57] Di Q, Xi C, Wang M, Zhao K, Li D, Wang F. Production of biodiesel using immobilized lipase and the characterization of different co-immobilizing agents and immobilization methods. Sustainability. 2016;**8**(9):764. DOI: 10.3390/su8090764

[58] Zaborsky O. Immobilized Enzymes. Cleveland: CRC Press; 1973. DOI: 10.1002/food.19750190519

[59] Zaborsky OR, Ogletree J. The immobilization of glucose oxidase via activation of its carbohydrate residues. Biochemical and Biophysical Research

Communications. 1974;**61**(1):210-216. DOI: 10.1016/0006-291X(74)90554-3

[60] Klibanov AM. Approaches to enzyme stabilization. Biochemical Society Transactions. 1983;**11**(1):19-20. DOI: 10.1042/bst0110019

[61] Germain P, Slagmolen T, Crichton RR. Relation between stabilization and rigidification of the three-dimensional structure of an enzyme. Biotechnology and Bioengineering. 1989;**33**(5):563-569. DOI: 10.1002/bit.260330508

[62] Brady D, Jordaan J. Advances in enzyme immobilisation. Biotechnology Letters. 2009;**31**(11):1639-1650. DOI: 10.1007/s10529-009-0076-4

[63] Coche-Guérente L, Cosnier S, Labbé P. Sol-gel derived composite materials for the construction of oxidase/peroxidase mediatorless biosensors. Chemistry of Materials. 1997;**9**(6):1348-1352. DOI: 10.1021/cm960498x

[64] Asgher M, Asad MJ, Bhatti HN, Legge RL. Hyperactivation and thermostabilization of Phanerochaete chrysosporium lignin peroxidase by immobilization in xerogels. World Journal of Microbiology & Biotechnology. 2007;**23**(4):525-531. DOI: 10.1007/s11274-006-9255-9

[65] Reetz MT. Review: Lipases as practical biocatalysts. Current Opinion in Chemical Biology [Internet]. 2002;**6**:145-150. DOI: 10.1016/S1367-5931(02)00297-1

[66] Reetz MT, Wenkel R, Avnir D. Entrapment of lipases in hydrophobic sol-gel-materials: Efficient heterogeneous biocatalysts in aqueous medium. Synthesis. 2000;**2000**(6): 781-783. DOI: 10.1055/s-2000-6276

[67] Wu JC, Selvam V, Teo HH, Chow Y, Talukder MMR, Choi WJ. Immobilization of *Candida rugosa* lipase by cross-linking with glutaraldehyde followed by entrapment in alginate beads. Biocatalysis and Biotransformation. 2006;**24**(5):352-357. DOI: 10.1080/10242420600787326

[68] Mauguet MC, Legrand J, Brujes L, Carnelle G, Larre C, Popineau Y. Gliadin matrices for microencapsulation processes by simple coacervation method. Journal of Microencapsulation. 2002;**19**(3):377-384. DOI: 10.1080/02652040110105346

[69] Moore CM, Akers NL, Hill AD, Johnson ZC, Minteer SD. Improving the environment for immobilized dehydrogenase enzymes by modifying nafion with tetraalkylammonium bromides. Biomacromolecules. 2004;**5**(4):1241-1247. DOI: 10.1021/bm0345256

[70] Iso M, Shirahase T, Hanamura S, Urushiyama S, Omi S. Immobilization of enzyme by microencapsulation and application of the encapsulated enzyme in the catalysis. Journal of Microencapsulation. 1989;**6**(2):165-176. DOI: 10.3109/02652048909098017

[71] Nguyen DT, Smit M, Dunn B, Zink JI. Stabilization of creatine kinase encapsulated in silicate sol-gel materials and unusual temperature effects on its activity. Chemistry of Materials. 2002;**14**(10):4300-4306. DOI: 10.1021/cm020398t

[72] Zhang Y, Wu H, Li J, Li L, Jiang Y, Jiang Y, et al. Protamine-templated biomimetic hybrid capsules: Efficient and stable carrier for enzyme encapsulation. Chemistry of Materials. 2008;**20**(3):1041-1048. DOI: 10.1021/cm701959e

[73] Tischer W, Kasche V. Immobilized enzymes: Crystals or carriers? Trends in Biotechnology. 1999;**17**(8):326-335. DOI: 10.1016/S0167-7799(99)01322-0

[74] Gonçalves MCP, Kieckbusch TG, Perna RF, Fujimoto JT, Morales S,

Romanelli JP. Trends on enzyme immobilization researches based on bibliometric analysis. Process Biochemistry. 2019;**76**:95-110. DOI: 10.1016/j.procbio.2018.09.016

[75] Migneault I, Dartiguenave C, Bertrand MJ, Waldron KC. Glutaraldehyde: Behavior in aqueous solution, reaction with proteins, and application to enzyme crosslinking. BioTechniques. 2004;**37**(5):790. DOI: 10.2144/04375RV01

[76] Garcia-Galan C, Fernandez-Lafuente R, Berenguer-Murcia A, Rodrigues RC. Potential of different enzyme immobilization strategies to improve enzyme performance. Advanced Synthesis and Catalysis. 2011;**353**(16):2885-2904. DOI: 10.1002/adsc.201100534

[77] St. Clair NL, Navia MA. Cross-linked enzyme crystals as robust biocatalysts. Journal of the American Chemical Society. 1992;**114**:7314-7316. DOI: 10.1021/ja00044a064

[78] Navia MA, Lexington M, St. Clair NL, Charlestown M. Crosslinked Enzyme Crystals. 19970408. Patent # 5,618,710; 1997

[79] Navia MA, Lexington M, St. Clair NL, Charlestown M. Method of protein therapy by orally administering crosslinked protein crystals. 20000104. Patent # 6,011,001; 2000

[80] Borman S. Cross-linked enzyme crystals show promise for industrial, clinical uses. Chemical & Engineering News. 1992;**70**:40-43. DOI: 10.1021/cen-v070n039.p040

[81] Häring D, Schreier P. Cross-linked enzyme crystals. Current Opinion in Chemical Biology. 1999;**3**:35-38. DOI: 10.1016/S1367-5931(99)80007-6

[82] Roy JJ, Abraham TE, Abhijith KS, Kumar PVS, Thakur MS. Biosensor for the determination of phenols based on cross-linked enzyme crystals (CLEC) of laccase. Biosensors & Bioelectronics. 2005;**21**(1):206-211. DOI: 10.1016/j.bios.2004.08.024

[83] Cao L, Van Rantwijk F, Sheldon RA. Cross-linked enzyme aggregates: A simple and effective method for the immobilization of penicillin acylase. Organic Letters. 2000;**2**(10):1361-1364. DOI: 10.1021/ol005593x

[84] Lopez-Serrano P, Cao L, van Rantwijk F, Sheldon RA. Cross-linked enzyme aggregates with enhanced activity: Application to lipases. Biotechnology Letters. 2002;**24**(16):1379-1383. DOI: 10.1023/A:1019863314646

[85] Sheldon RA, Schoevaart R, van Langen LM. Cross-linked enzyme aggregates. In: Guisan JM, editors. Immobilization of Enzymes and Cells. Methods in Biotechnology™. Vol. 22. Totowa, New Jersey: Humana Press; 2006. pp. 31-46. DOI: 10.1007/978-1-59745-053-9_3

[86] Goodford PJ. A computational procedure for determining energetically Favorable binding sites on biologically important macromolecules. Journal of Medicinal Chemistry. 1985;**28**(7):849-857. DOI: 10.1021/jm00145a002

[87] Cao S, Xu P, Ma Y, Yao X, Yao Y, Zong M, et al. Minireview: Recent advances in immobilized enzymes on nanocarriers. Chinese Journal of Catalysis. 2016;**37**:1814-1823. DOI: 10.1016/S1872-2067(16)62528-7

[88] Kim J, Jia H, Wang P. Challenges in biocatalysis for enzyme-based biofuel cells. Biotechnology Advances. 2006;**24**(3):296-308. DOI: 10.1016/j.biotechadv.2005.11.006

[89] Yiu HHP, Keane MA. Enzyme-magnetic nanoparticle hybrids:

New effective catalysts for the production of high value chemicals. Journal of Chemical Technology and Biotechnology. 2012;**87**(5):583-594. DOI: 10.1002/jctb.3735

[90] Pal A, Khanum F. Covalent immobilization of xylanase on glutaraldehyde activated alginate beads using response surface methodology: Characterization of immobilized enzyme. Process Biochemistry. 2011;**46**(6):1315-1322. DOI: 10.1016/j.procbio.2011.02.024

[91] Cedrone F, Ménez A, Quéméneur E. Tailoring new enzyme functions by rational redesign. Current Opinion Structural Biology. 2000;**10**:405-410

[92] Arnold FH. Combinatorial and computational challenges for biocatalyst design. Nature. 2001;**409**(6817):253-257. DOI: 10.1038/35051731

[93] Bornscheuer UT, Huisman GW, Kazlauskas RJ, Lutz S, Moore JC, Robins K. Engineering the third wave of biocatalysis. Nature. 2012;**485**(7397): 185-194. DOI: 10.1038/nature11117

[94] Wiegand K, Farès C, Kannchen D, Rögner M, Winkler M, Rexroth S, et al. Rational redesign of the ferredoxin-NADP+-oxido-reductase/ferredoxin-interaction for photosynthesis-dependent H2-production. Biochimica et Biophysica Acta. 2018;**1859**:253-262

[95] Antikainen NM, Martin SF. Review: Altering protein specificity: Techniques and applications. Bioorganic & Medicinal Chemistry. 2005;**13**:2701-2716. DOI: 10.1016/j.bmc.2005.01.059

[96] Liu H, Naismith JH. An efficient one-step site-directed deletion, insertion, single and multiple-site plasmid mutagenesis protocol. BMC Biotechnology. 2008;**8**:91. DOI: 10.1186/1472-6750-8-91

[97] Yang H, Liu L, Shin H, Chen RR, Li J, Du G, et al. Structure-based engineering of histidine residues in the catalytic domain of α-amylase from *Bacillus subtilis* for improved protein stability and catalytic efficiency under acidic conditions. Journal of Biotechnology. 2013;**164**:59-66. DOI: 10.1016/j.jbiotec.2012.12.007

[98] Duan X, Cheng S, Ai Y, Wu J. Enhancing the thermostability of serratia plymuthica sucrose isomerase using B-factor-directed mutagenesis. PLoS One. 2016;**11**(2):1-16. DOI: 10.1371/journal.pone.0149208

[99] Chen R. Enzyme engineering: Rational redesign versus directed evolution. Trends in Biotechnology. 2001;**19**(1):13-14. DOI: 10.1016/S0167-7799(00)01522-5

[100] Arnold FH, Volkov AA. Directed evolution of biocatalysts. Current Opinion in Chemical Biology. 1999;**3**(1):54-59. DOI: 10.1016/s1367-5931(99)80010-6

[101] Labrou NE. Random mutagenesis methods for in vitro directed enzyme evolution. Current Protein and Peptide Science. 2010;**11**(1):91-100. DOI: 10.2174/138920310790274617

[102] Reetz MT. Changing the enantioselectivity of enzymes by directed evolution. Methods in Enzymology. 2004;**388**:238-256. DOI: 10.1016/S0076-6879(04)88021-2

[103] Ness JE, Del Cardayré SB, Minshull J, Stemmer WPC. Molecular breeding: The natural approach to protein design. Advances in Protein Chemistry. 2001;**55**:261-292. DOI: 10.1016/S0065-3233(01)55006-8

[104] Lutz S. Beyond directed evolution—Semi-rational protein engineering and design. Current Opinion in Biotechnology. 2010;**21**:734-743. DOI: 10.1016/j.copbio.2010.08.011

[105] Garcia-Ruiz E, Gonzalez-Perez D, Ruiz-Duenas FJ, Martinez AT, Alcalde M. Directed evolution of a temperature-, peroxide- and alkaline pH-tolerant versatile peroxidase. Biochemical Journal. 2012;**441**:487-498. DOI: 10.1042/BJ20111199

[106] Alvizo O, Nguyen L, Savile CK, Bresson JA, Lakhapatri SL, Solis E, et al. Directed evolution of an ultrastable carbonic anhydrase for highly efficient carbon capture from flue gas. Proceedings of the National Academy of Sciences of the United States of America. 2014;**111**(46):16436. DOI: 10.1073/pnas.1411461111

[107] Chao F-A, Morelli A, Iii JCH, Churchfield L, Hagmann LN, Masterson LR, et al. Structure and dynamics of a primordial catalytic fold generated by in vitro evolution. Nature Chemical Biology. 2013;**9**(2):81-83. DOI: 10.1038/nchembio.1138

[108] Reetz MT, Carballeira JD, Peyralans J, Hoebenreich H, Maichele A, Vogel A. Expanding the substrate scope of enzymes: Combining mutations obtained by CASTing. Chemistry—A European Journal. 2006;**12**(23):6031-6038. DOI: 10.1002/chem.200600459

[109] Steiner K, Schwab H. Recent advances in rational approaches for enzyme engineering. Computational and Structural Biotechnology Journal. 2012;**2**(3). Article ID: e201209010. DOI: 10.5936/csbj.201209010

[110] Roth T, Beer B, Pick A, Sieber V. Thermostabilization of the uronate dehydrogenase from agrobacterium tumefaciens by semi-rational design. AMB Express. 2017;**7**(1):1-9. DOI: 10.1186/s13568-017-0405-2

[111] Röthlisberger D, Khersonsky O, Wollacott AM, Jiang L, DeChancie J, Betker J, et al. Kemp elimination catalysts by computational enzyme design. Nature. 2008;**453**(7192): 190-195. DOI: 10.1038/nature06879

Chapter 4

Study of the Influence of Humic Acid Macromolecules on the Structure of Erythrocytes of Some Animals by the Method of Absorption

Nikolay L. Lavrik and Tatiana N. Ilyitcheva

Abstract

Erythrocyte absorption spectra were obtained from fresh chicken, goose, and guinea pig blood in solutions with humic acids, isolated from brown coal, to study interactions between erythrocytes and humic acids (HA). It has been established that the addition of HA to erythrocytes leads to the differently directed shifts of Soret band maxima in the erythrocyte absorption spectrum. Thus, for a solution [guinea pig erythrocyte (1.5×10^{12} particle/l) + HA №1], this difference was $\Delta\lambda = +3.3$ nm (shortwave shift); for a solution [chicken erythrocyte (2×10^{12} particle/l) + HA №1], $\Delta\lambda = -1.5$ nm (longwave shift); and for a solution [goose erythrocyte (6×10^{11} particle/l) + HA №1], $\Delta\lambda = +4.3$ nm (shortwave shift). A comparison of the absorption spectra of guinea pig oxyhemoglobin with 2 HA samples indicates that at any erythrocyte concentrations, the positions of the Soret band maxima for various HA samples differ. The data obtained testify to the individual character of the interaction between erythrocyte membranes and HA macromolecules. Two hypotheses were proposed to account for the results obtained. (1) "Structural hypothesis." In the framework of this hypothesis, the molecules of membrane-bound oxyhemoglobin are in erythrocyte volume and can undergo noticeable, structural changes due to the deformation of erythrocyte membrane. (2) "Complexing hypothesis." In terms of this hypothesis, the observed shifts of the position of the Soret band maxima can be explained by the possible penetration of light HA fragments through erythrocyte membrane into the inner erythrocyte region. This can cause the formation of complexes (oxyhemoglobin-HA). In this case, the complex formation can involve both the free oxyhemoglobin molecules (HbO_2) and the membrane-bound ones.

Keywords: chicken, goose, guinea pig blood erythrocytes, humic acids, absorption spectra, Soret band

1. Introduction

Recently, an increase has been observed in the interest to the application of humic acids (HA) in medicine. These are the examples of HA effect on some vitally important properties of human organism [1]:

1. Antioxidant properties. A humic complex manifests an explicit ability to support chemical balance in organism. Depending on the situation, humic acid can behave itself either as donor or electron acceptor. This makes humic acid a powerful, natural antioxidant, the trap of free radicals that damage protein structures and DNA molecules of cells, break their genetic code, and, in particular, promote the development of oncological diseases.

2. Antiviral activity. Humic acids exhibit a high antiviral activity. A humic acid molecule covers a virus as a "coat" to block its escape into the bulk and prevents its reproduction. In this case, humic acid sends a signal to immune system about the appearance of an invader. This pushes the immune system to fight the virus which is in a vulnerable position (bound to a humic acid molecule). As a result, the number of viruses decreases, and the immune system successfully fights the disease.

3. Detoxicant and hepatoprotector. Humic acids are a powerful means of complexing. They bind and remove heavy metals (lead, copper, mercury, cadmium, cobalt, zinc, etc.) from the body. At a certain concentration, these cause severe poisoning and cell mutations. Heavy metals are not removed independently without special therapeutic measures. Humic acids participate actively in liver metabolism and act as a filter for heavy metals. They capture and immobilize toxic substances, preventing them from taking part in chemical reactions. Thereafter, toxin is readily removed from the body.

4. Influence on blood properties. Humic acids, in the amount of 100–300 mg per kg, have no effect on the time of bleeding, the time of blood cluttering, thrombin time, or platelet aggregation. Red cells and hemoglobin remain at normal levels. In this case, in the presence of humic acids, erythrocytes carry a higher percentage of oxygen to tissues.

5. Antibacterial activity. Humic acids have a pronounced antibacterial action on the following pathogenic microorganisms: *C. albicans*, *P. vulgaris*, *P. aeruginosa*, *S. typhimurium*, *S. aureus*, *S. epidermidis*, and *S. pyogenes*. They substantially accelerate bacterium metabolism which leads to a strong destruction of bacterial cells. In an intestinal tract, humic acids neutralize a pathogenic microflora. The bound bacteria and toxins are excreted naturally.

6. Immune system. One of the most pronounced HA effects is strengthening of the general immune response. Humic compounds regulate the number of glycoproteins that affect the balance of T- and B-lymphocytes. In addition, these activate the synthesis of interleukins 1 and 2 and the production of endogenous interferon and gamma-globulins which activates the oppressed functions of immune system. A series of clinical investigations indicate that humic acids can manifest the anticancer properties by inhibiting tumor growth and suppressing the action of viruses that can cause the development of cancer. Fulvic acid decreases protease activity which allows one to decrease the metastatic activity of cancer cells.

7. Anti-inflammatory properties. Humic acids have an anti-inflammatory action. These accelerate the healing of wounds and ulcerative defect by strengthening the processes of fibroblast proliferation, acceleration of water, protein, and lipid exchange. They also inhibit the synthesis of inflammation mediators—prostaglandins. The tissue hyaluronidase, which accelerates wound healing, is

activated locally. Humic acids were established to inhibit proteolytic enzymes that damage the walls of vessels and skin.

8. Antiatherosclerotic effect. Since humic acids can distinguish and bind substances, present in excess in the body, they form and remove the complexes with cholesterol and lipoproteins of low density which makes them efficient in their fight against atherosclerosis and its effects.

9. Antiallergic effect. Humic acids decrease organism sensibilization by actively binding and removing allergens from organisms. In this case, the symptoms of allergy vanish, the number of eosinophils in blood is normalized, and stable remission is attained.

10. Anti-stress effect. Humic acids regulate the action of stress hormones produced by adrenals (adrenaline, noradrenaline). The high level of adrenaline and noradrenaline indicates an increased level of anxiety. Excess hormones are blocked by humic acids and fail to reach their receptors in a cell. In addition, the ability of humic acids to affect the saturation of red blood cells with oxygen improves overall health and causes a surge of strength.

On humic acid structure, humic acid macromolecules are the polymer of variable molecular size and composition [2, 3]. At present, there is no full clarity in the understanding of a concrete structure of HA macromolecules, and only the general structural peculiarities are available. In terms of the generally accepted concepts, chemically, humic acids are the highly molecular nitrogen-containing organic acids whose molecules include aromatic groupings. A general pattern of the structure of HA macromolecule is as follows. There is a nucleus (aromatic carbon skeleton) and a periphery (polysaccharide-polypeptide chains) ([4]; **Figure 1**). It is assumed then that the molecular fragments of the nucleus and periphery of one HA macromolecule are bound by chemical bonds. The condensed aromatic nuclei, bound by the chains with a fair conjugation of carbon-carbon and other bonds, are the carriers of the specific properties of humic acids. Peripheral, irregular, structural elements (peripheral chains) are the variable components. As a result, the structure of HA macromolecules is unstable and subjected to statistic fluctuations. As a whole, the

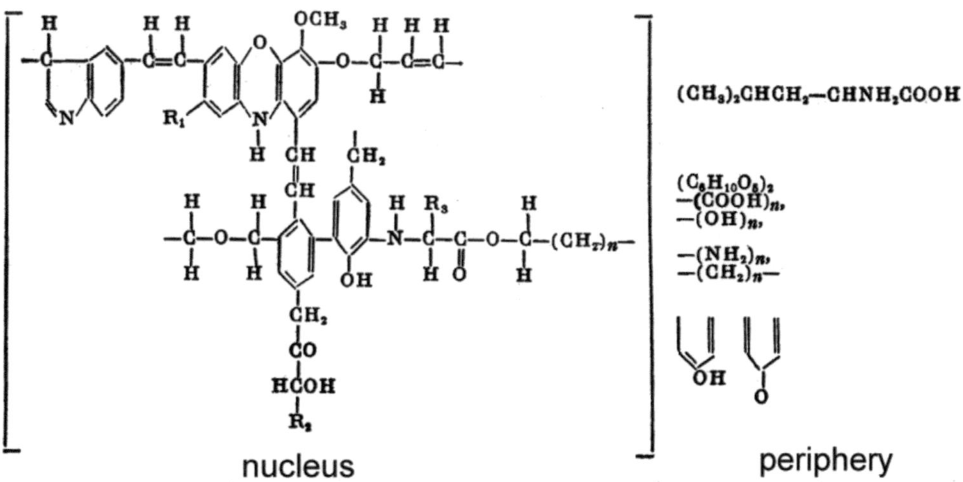

Figure 1.
Model of macromolecule structure of humic acid according to Felbeck [4].

HA macromolecules are characterized by statistically continuous range of various structural units. Thus, a characteristic feature of HA macromolecules is their polydispersion.

Recently, along with the traditional viewpoint on the structure of HA macromolecules, there appeared the alternative one [5]. In the framework of the alternative concept, the HA macromolecule structure is the supramolecular self-organizing ensemble of heterogeneous and relatively small molecules, arising from a dead biological material, rather than a single molecule in which the various structural fragments are bound by covalent bonds. The most important property of such a humic, supramolecular structure is that it is stabilized not by covalent bonds but by weak dispersion forces (van der Waals, π-π interactions, and CH-π interactions) and by H-bonds. The efficiency of HA macromolecule complexing with various simple organic and inorganic compounds is known, at present, in detail [6–16]. It is worth noting that all the works on the study of HA complexing were performed without the method of absorption and information on HA interaction with erythrocytes is unavailable at all.

It is concluded then that by now, there are numerous data on the effect of HA on the vitally essential functions of human organisms. However, the medical HA-based preparations (HA preparations) are not at present widely used in medicine.

One of the reasons, constraining a broad development and implementation of HA preparations, is the absence of a systematic study and, thus, the absence of fundamental knowledge of the mechanism of interaction between humic substance and cell at the molecular level. In particular, there are no data on the interaction of HA macromolecules with such an important cell, contained in blood, as erythrocyte. Information about the HA-erythrocyte interaction can be extracted by observing a supramolecular effect, i.e., the presence of hemagglutination.

It is known that humic substances can agglutinate erythrocytes. As verified by our preliminary experiments, the HA components, obtained from brown coal, selectively agglutinate human and animal erythrocytes: some components agglutinate human and chicken erythrocytes but fail to agglutinate the goose and guinea pig ones. On the contrary, the other types of humic acids agglutinated the goose and guinea pig erythrocytes and failed in the case of chicken and human erythrocytes. Finally, some types of HA agglutinated all erythrocytes used in experiments, whereas the other types of humic acids could not agglutinate erythrocytes at all. In particular, the samples of commercial, artificial fulvic and humic acids did not agglutinate erythrocytes. These results indicate that upon agglutination, the specific HA-cell interactions occur that allow one to distinguish their binding sites; i.e., the interaction efficiency is individual for a concrete HA-erythrocyte pair. Thus, of interest is the study on the efficiency of HA-erythrocyte interaction at the molecular level.

The goal of this work was to determine the efficiency of the interaction between erythrocytes of some animals and HA macromolecules by the method of absorption. Information on the efficiency of HA-erythrocyte interaction can be extracted from the experiments on the change in the parameters of Soret absorption band, that, as shown in [17], are sensitive to the change in the state of oxyhemoglobin molecule, HbO_2, contained in erythrocytes. The visible region of HbO_2 absorption spectra exhibits the three strongest characteristic bands with maxima at $\lambda \sim 415$ nm (Soret band), $\lambda \sim 545$ nm (β band), and $\lambda \sim 580$ nm (α band) [18, 19]. The origin of HbO_2 absorption spectra has been established quite reliably. It is defined by the electron properties of hemes that are the prosthetic groups in the structure of hemoglobins [20]. In metalloporphyrins, the Soret band (heme molecules in erythrocytes) is determined by both the electronic $\pi \rightarrow \pi^*$ transition $2^1E_a \leftarrow 1^1A_g$, when the upper state is twice degenerated (D_{4h} symmetry), and the electronic

$\pi \rightarrow \pi^*$ transitions $2^1B_{3u} \leftarrow 1^1A_g$ and $2^1B_{2u} \leftarrow 1^1A_g$ in the absence of degeneracy (D_{2h} symmetry) [19, 20]. Experimentally, the Soret spectrum of oxyhemoglobin in erythrocytes exhibits one band which allows us to assign the origin of the spectrum observed to the electron transition $2^1E_a \leftarrow 1^1A_g$.

On the possibility to observe spectral changes in the Cope band spectrum of oxyhemoglobin molecule upon HA addition.

The changes in the HbO_2 absorption spectrum upon both HA addition and conservation of initial erythrocyte can be a priori expected due to the following. In the initial erythrocytes, about 10% of HbO_2 can be bound to the inner surface of the membrane [21–24], i.e., the membrane-bound oxyhemoglobin results from the interaction between HbO_2 molecule and membrane components (band 3 protein, spectrin, glycophorin, membrane lipids). The form of erythrocytes can vary after addition of HA and upon hemagglutination which can have effect on the structure of the membrane-bound HbO_2 molecules, i.e., cause changes in heme structure. Thus, the structural rearrangement of the membrane-bound HbO_2 molecules can lead to the changes in their spectral parameters. As a result, the absorption spectra of initial erythrocyte samples and of those that can agglutinate will be different. The position of absorption band maxima depends on the state of the electron orbitals of porphyrin rings, contained in the heme. The sensitivity of electron transitions between orbitals to the structural changes in porphyrin rings, which form a heme, has been established quite reliably [19, 20]. The Soret band is determined by the electron $\pi \rightarrow \pi^*$ transition which is highly sensitive to the change in structure and environment [25]. In particular, as has been established, upon binding of porphyrins to DNA, the Soret band undergoes a long-wave shift of up to 20 nm upon internal intercalation and of 8 nm upon the external one [26]. Besides, the addition of protein causes a bathochromic shift of 10 nm in the chlorine absorption spectrum, contained in porphyrin rings [27]. At last, the effect of erythrocyte lysis on the position of the Soret band maxima is reported in [17]. Thus, the spectral parameters of the oxyhemoglobin molecule absorption spectrum are assumed to change upon hemagglutination due to the deformation of erythrocyte structure.

The goose, chicken, and guinea pig erythrocytes of different geometric parameters and the HA, isolated from brown coal, were used as the samples for solving the problem stated. A preliminary work indicates that the HA preparation agglutinates the chicken and guinea pig erythrocytes and does not agglutinate the goose ones.

2. Experimental

Isolation of humic acids. In this work, we used humic acids isolated from brown coal by two different techniques. The first one [28] includes the treatment of coal with alkaline reagents with subsequent oxidation in a cathode chamber of a diaphragm electrolyzer at room temperature and atmospheric pressure. In the second technique [29], brown coal was treated with an aqueous ammonia solution of 5–25% upon mixing. Cyclohexanol, or its analog, was then added in the amount of 0.1–1.0% of the initial coal mass. Hydrogen peroxide was added in small portions by stirring until the maximal dissolution of brown coal. The concentration of NaCl in HA solutions amounted to 0.9% (a standard physiological solution (0.9% NaCl "Fluka") was used). pH of working solutions was 7.2 (NaOH).

Production of erythrocytes. To prepare erythrocytes, the samples of fresh chicken, goose, or guinea pig blood with 5–10 units/ml of heparin were used and then filtered through a sterile cotton gauze cloth (Fisher, cat. № 22-415-469) into a conical 500 ml tube which was then carefully filled with cold 0.01 M phosphate-saline buffer (PSB, pH = 7.2), capped and stirred slightly while turning the tube. Centrifugation was performed at 1200 rpm/min for 10 min at +4°C. The supernatant was then removed using a 10 ml pipette. This was followed by a cold phosphate-saline buffered double wash. The remaining supernatant was removed with a micropipette. Compacted erythrocytes were stored on ice. The erythrocyte suspension (1%) was prepared by adding 2.5 ml of compacted erythrocytes to 247.5 ml of cold PSB in a 50 ml glass flask and stirred via rotation. The concentrations of erythrocytes from guinea pig, chicken, and goose blood were 30 × 10^{12}, 40 × 10^{12}, and 12 × 10^{12} particle/l, respectively.

Preparation of solutions of erythrocytes with humic acids. The concentrations of erythrocyte solutions in HA were prepared as follows. The initial solutions were diluted 10 times. 0.3, 0.5, 1, and 1.5 ml were poured into a tube, and the HA solution was added to obtain 10 ml.

Obtaining absorption spectra. The absorption spectra were recorded on a Hewlett Packard 6041 spectrophotometer. Standard 1 cm quartz cuvettes were used.

Determination of absorption band maximum. The position of the absorption band maxima was determined by taking the first absorption spectrum derivative. The value of the abscissa at which the first derivative was zero was taken as the position of a maximum. The absorption spectrum derivative was determined using a standard method, contained in the program "Origin 7."

3. Results and discussion

Figure 2 shows the absorption spectra of erythrocyte samples of various concentrations in HA solution, isolated from the guinea pig (a), chicken (b), and goose (c) blood. In the HA solutions, the obtained spectra of the initial erythrocyte samples also coincide with the available data on oxyhemoglobin, whose spectra exhibit the bands at λ ~ 415 nm, λ ~ 545 nm, and λ ~580 nm [18, 19]. The presence of characteristic bands at λ ~ 545 nm and at λ~580 nm in the spectra of erythrocyte samples with HA indicates that the HA solutions contain just the oxyhemoglobin molecules because this doublet is not recorded for other hemoglobin derivatives (deoxyhemoglobin, methemoglobin, carboxyhemoglobin, and hemichrome) [18, 19, 30].

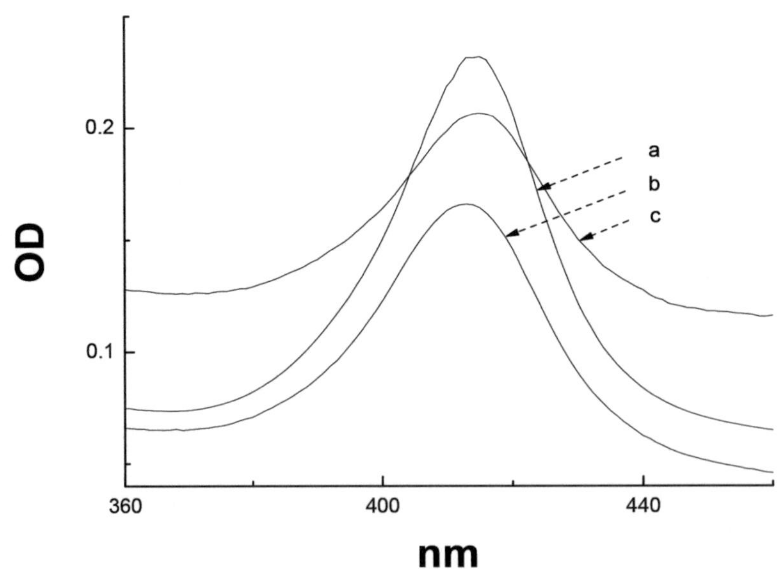

Figure 2.
Absorption spectra of erythrocyte samples. (a) guinea pig, [Erythr] = 3 × 10^{12} particles/k; (b) chicken, [Erythr] = 4 × 10^{12} particles/l; and (c) goose, [Erythr] = 1.2 × 10^{12} particles/l.

Figure 3 demonstrates the absorption spectra of oxyhemoglobin of different concentrations in HA solutions. According to the data presented, as the concentration of erythrocytes decreases, the position of the Soret band maximum undergoes a noticeable long-wave shift for goose and chicken erythrocytes, and in the case of the guinea pig ones, the shift is very weak. Thus, the interaction between erythrocytes and HA causes the maximal spectral changes in goose erythrocytes, and the minimal ones are observed in the chicken erythrocytes.

The highest effect of the difference in the position of the Soret $\Delta\lambda$ band maximum is observed by comparing the absorption spectra of the initial sample erythrocytes with the sample of (erythrocyte + HA} solution. Thus, for a solution [guinea pig erythrocyte (1.5×10^{12} particle/l) + HA №1], this difference was $\Delta\lambda$ = +3.3 nm (shortwave

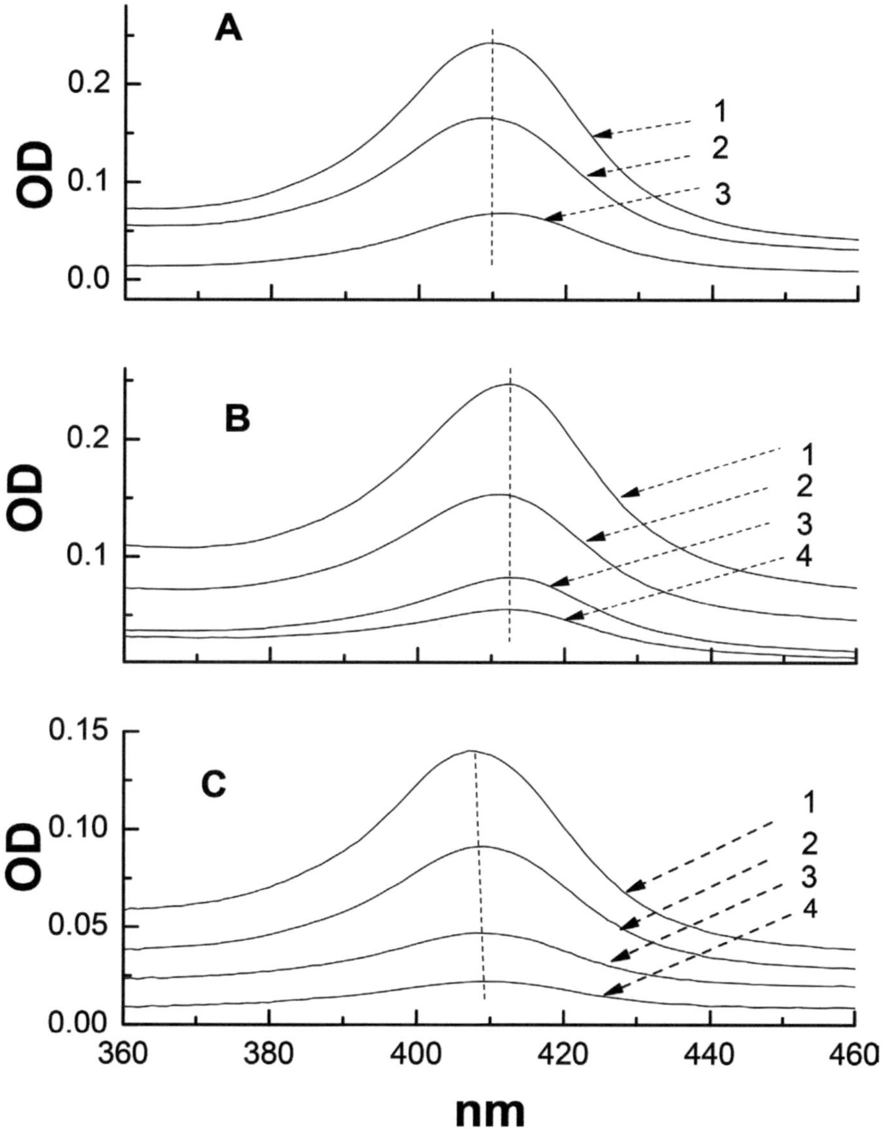

Figure 3.
Absorption spectra of erythrocytes of different concentrations in HA solutions. (A) guinea pig: (1) [Erythr] = 4.5×10^{11} particles/l, (2) [Erythr] = 3×10^{11} particles/l, and (3) [Erythr] = [1.5×10^{11} particles/l; (B) chicken: (1) [Erythr] = 6×10^{11} particles/l, (2) [Erythr] = 4×10^{11} particles/l, (3) [Erythr] = 2×10^{11} particles/l, and (4) [Erythr] = 1.2×10^{11} particles/l; and (C) goose, (1) [Erythr] = 1.8×10^{12} particles/l., (2) [Erythr] = 1.2×10^{11} particles/l, (3) [Erythr] = 0.6×10^{12} particles/l, and (4) [Erythr] = 0.36×10^{11} particles/l.

shift); for a solution [chicken erythrocyte (2×10^{12} particle/l) + HA №1], $\Delta\lambda = -1.5$ nm (longwave shift); and for a solution [goose erythrocyte (6×10^{11} particle/l) + HA№1], $\Delta\lambda = +4.3$ nm (shortwave shift). Thus, there is a difference in the change in the position of the Soret band maxima by both the absolute value and the direction.

However, a conclusion can be now drawn on the individual character of the effect of HA interaction with the erythrocytes of the animals studied, observed at the molecular level, which correlates with the previous information on the individual character of the hemagglutination of the erythrocytes of these animals.

Figure 4 shows the absorption spectra of guinea pig oxyhemoglobin of various concentrations with HA samples №1 and №2. As follows from the figure, at any erythrocyte concentration, the positions of the Soret band maxima for HA sample №1 are always in a redder region of the spectrum. Thus, the efficiency of the interaction between the guinea pig erythrocytes and HA samples №1 and №2 is different. This effect is, probably, due to the difference in the structural properties of HA samples (**Figure 5**).

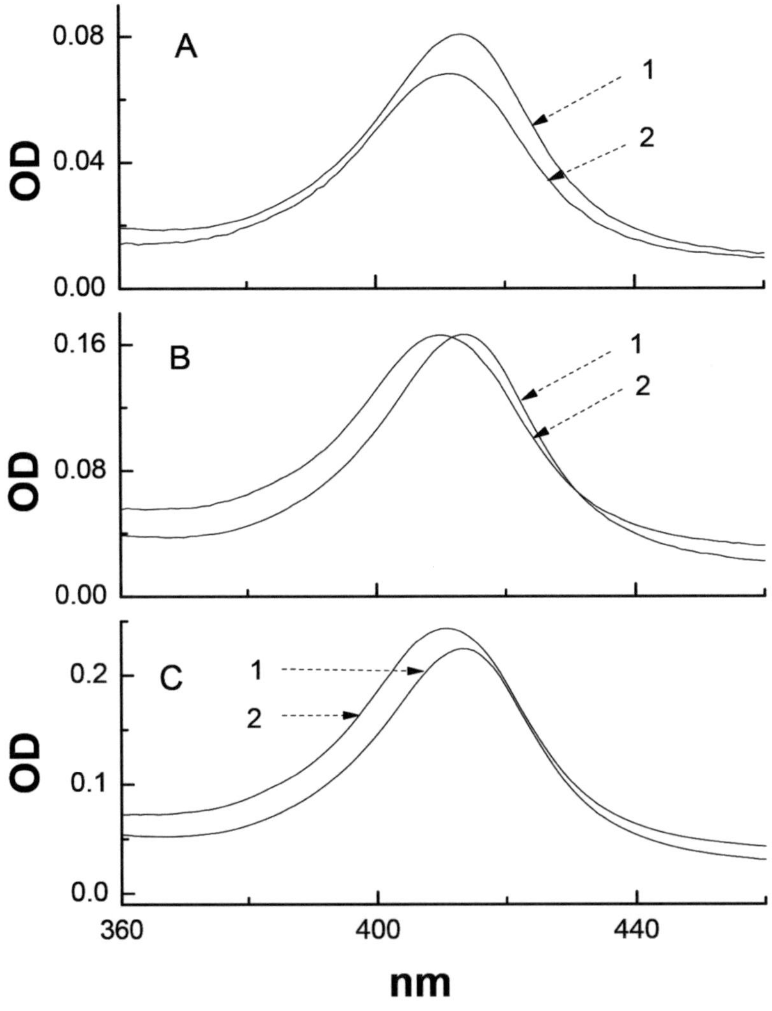

Figure 4.
Absorption spectra of guinea pig oxyhemoglobin of different concentrations with HA samples №1 and №2. (A) [Erythr] = 1.5×10^{11} particle/l, (B) [Erythr] = 3×10^{11} particle/l, and (C) [Erythr] = 4.5×10^{11} particle/l. (1) HA №2 and (2) HA №1.

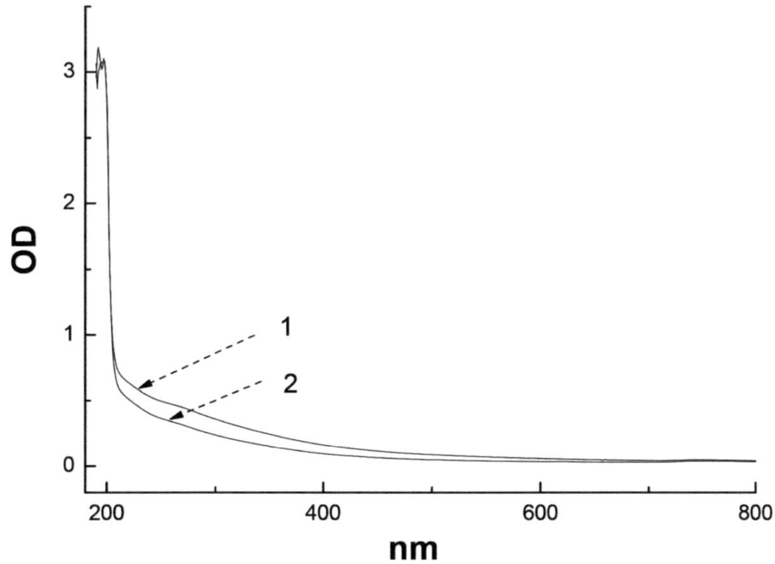

Figure 5.
HA absorption spectra [(2.5 mg/l)]. (1) №1 and (2) №2.

4. Discussion of results

1. "Structural hypothesis." As mentioned in Introduction, the addition of HA and the presence of hemagglutination cause changes in erythrocyte form which then can have an effect on the structure of the membrane-bound HbO_2 molecules, i.e., finally, can lead to the change in heme structure. Thus, the spectral parameters of the membrane-bound HbO_2 molecules can change due to their structural rearrangement. As a result, the absorption spectra of the initial erythrocyte samples without addition of HA and erythrocytes that agglutinate will differ. It is readily seen that in the framework of this hypothesis, the molecules of the membrane-bound oxyhemoglobin must undergo noticeable structural changes due to the deformation of erythrocyte membrane. The possibility of the deformation of erythrocyte structure upon interaction with other molecules was reported, e.g., in [31]. In this work, the method of atomic force microscopy was used to verify that hemin has a specific effect on the nanostructure of erythrocyte membranes by forming domains on its surface.

2. "Complexing hypothesis." The shift observed in the position of the Soret band maximum can be differently explained by a feasible penetration of light HA fragments through erythrocyte membrane into the inner erythrocyte region. (This mechanism can be similar to the penetration of virus into a cell. The possibility of this process was verified by studies on the interaction between human erythrocytes and carnosine molecule [32]. In any sample of HA, due to the property of polydispersity of HA [2], light fragments are always present. Direct evidence of the presence of light fractions in the "Aldrich" sample was shown in [33]). The penetration of HA fragments into erythrocyte can lead to the formation of oxyhemoglobin-HA complexes. In this case, this process can involve both the free oxyhemoglobin molecules and the membrane-bound ones. However,

this hypothesis fails to account for the experimentally observed shortwave shift of the Soret band with increasing erythrocyte concentration, Soret band origin and $\pi^* \rightarrow \pi$ transitions, and the process of complexing must cause a long-wave shift [25].

It is impossible now to give preference to one of these hypotheses.

Author details

Nikolay L. Lavrik[1]* and Tatiana N. Ilyitcheva[2]

1 Voevodsky Institute of Chemical Kinetics and Combustion, SB RAS, Novosibirsk, Russia

2 State Scientific Center of Virology and Biotechnology "Vektor", Koltzovo, Novosibirsk, Russia

*Address all correspondence to: lavrik@kinetics.nsc.ru

References

[1] Humic Acids are the Missing Link in the Food Chain: Available from: https://econet.ru/articles/173071

[2] Stevenson FJ. Humus Chemistry. Genesis, Composition, Reactions. 2nd ed. New York: John Wiley and Sons, Inc; 1994. 386 p

[3] Orlov DS. Humic Acids and General Theory of Humification. Moscow: MGU; 1990. p. 325

[4] Felbeck GT. Structural chemistry of soil humic substances. Advances of Agronomy. 1965;**17**:327-368

[5] Piccolo A. The supramolecular structure of humic substances: A novel understanding of humic chemistry and implications in soil science. Advances in Agronomy. 2002;**75**:57-134

[6] Lavrik NL, Mulloev NU. Observation of the formation of humic acid complexes with copper ions by absorption. Zhurnal Prikladnoi Spektroskopii. 2014;**81**:159-161

[7] Chen S, Inskeep W, Williams S, Calls P. Fluorescence lifetime measurements of fluoranthene, 1-naphthol, and napropamide in the presence of dissolved humic acid. Environmental Science and Technology. 1994;**28**:1582-1588

[8] Danielsen KM, Yu C, Butterbauch J, Gustafson T, Traina S. Solubility enhancement and fluorescence quenching of pyrene by humic substances. Environmental Science and Technology. 1995;**29**:2162-2165

[9] Korshin GV, Frenkel AI, Stern EA. EXAFS study of the inner shell structure in copper (II) complexes with humic substances. Environmental Science & Technology. 1998;**32**(18):2699-2705

[10] Monteil-Rivera F, Dumonceau J. Fluorescence spectrometry for quantitative characterization of cobalt (II) complexation by Leonardite humic acid. Analytical and Bioanalytical Chemistry. 2008;**374**:1105-1109

[11] William T, Bresnahan C, Grant L, Weber JH. Stability constants for the complexation of copper(II) ions with water and soil fulvic acids measured by an ion selective electrode. Analytical Chemistry. 1978;**50**(12):1675-1680

[12] Gauthler TD, Seltz WR, Grant CL. Effect of structural and compositional variations of dissolved humic materials on pyrene K_{oc} values. Environmental Science & Technology. 1987;**21**(3):243-248

[13] Ma X, Green SA. Fractionation and spectroscopic properties of fulvic acid and its extract. Chemosphere. 2008;**72**:1425-1434

[14] Yashchenko NY, Perminova IV, Petrosyan VS, Fillipova EM, Fadeev VV. Interaction of humic acids of different origin with polyaromatic hydrocarbons: Effect of pH and ion force of solution. Vestnik MGU, Khimiya. 1999;**40**(2): 188-193

[15] Ryan DK, Weber JH. Copper (II) complexing capacities of natural waters by fluorescence quenching. Environmental Science & Technology. 1982;**6**(12):866-872

[16] Piana MJ, Zahir KO. Investigation of metal ions binding of humic substances using fluorescence emission and synchronous-scan spectroscopy. Journal of Environmental Science and Health, Part B. 2000;**35**(1):87-102

[17] Lavrik NL, Il'icheva TN. The effect of erythrocyte Lysis in selected animals on the absorption spectra of oxyhemoglobin. Biophysics. 2018;**63**(5):718-726

[18] Cantor CR, Schimmel PR. Biophysical Chemistry. San Francisco: Freeman; 1980

[19] Solovyev KN, Gladkov LL, Starukhin AS. Spectroscopy of Porphyrins: Oscillatory State. Minsk: Nauka i Tekhnika; 1985

[20] Sergeeva TN, editor. Porphyrins: Spectroscopy, Electrochemistry, Applications. Moscow: Nauka; 1987

[21] Shaklai N, Yguerabide J, Ranney HM. Classification and localization of hemoglobin binding sites on the red blood cell membrane. Biochemistry. 1977;**16**(25):5593-5599

[22] Walder JA, Chatterjee R, Steck TL. The interaction of hemoglobin with the cytoplasmic domain of band 3 of the human erythrocyte membrane. The Journal of Biological Chemistry. 1984;**259**(16):10238-10246

[23] Schluter K, Drenckhahn D. Co-clustering of denatured hemoglobin with band 3: Its role in binding of autoantibodies against band 3 to abnormal and aged erythrocytes. Proceedings of the National Academy of Sciences of the United States of America. 1986;**83**:6137-6145

[24] Datta P, Chakrabarty S, Chakrabarty A. Membrane interactions of hemoglobin variants, HbA, HbE, HbF and globin subunits of HbA: Effects of aminophospholipids and cholesterol. Biochimica et Biophysica Acta. 2008;**1778**(1):1-9

[25] Parker CA. Photoluminescence of Solutions: With Applications to Photochemistry and Analytical Chemistry. Mir, Moscow: Elsevier; 1968. 1972

[26] Pasternack RF, Gibbs EJ, Villafranca JJ. Interactions of porphyrins with nucleic acids. Biochemistry. 1983;**22**(10):2406-2411

[27] Golovina GV. Complex formation of tetra pyrrol compounds with albumin and lipoproteins. Candidate's Dissertation in Chemistry. Moscow; 2014

[28] Miloshenko TP, Kolyagin GA, Shklyaev AA, Shchipko ML, Kuznetsov BN, Kornienko VL. Authors's Certificate SSSR N 13938501988. pp. 57-60

[29] Miloshenko TP, Golovin YG, Golovina VV, Eremina AO, Mikhailenko CA. Method to obtain liquid water-soluble brown coal products Application: 93034073/04, 01.07.1993 Published: 27.09.1995

[30] Nikitin VN. Atlas of Blood Cells of Agricultural and Laboratory Animals. Moscow: Gos. Izd. S-kh. Lit; 1949

[31] Moroz VV, Kozlova EK, Chernysh AM, Gudkova OE, Bushueva AV. Changes in the structure of erythrocyte membranes under hemin action. General Reanimathology. 2012;**2**(6):5-10

[32] Arzumanyan EC. Protective action of carnosine on the hemolysis of erythrocytes accelerated by homocysteine acid. New Medical Preparations. 2007;**9**:12-15

[33] Song J, Huang W, Peng P, Ma BXY. Humic acid molecular weight estimation by high-performance size-exclusion chromatography with ultraviolet absorbance detection and refractive index detection. Soil Science Society of America Journal. 2010;**74**:2013-2020

Section 2

Therapeutic Applications

Chapter 5

Molecular Target Therapy against Neuroblastoma

Hidemi Toyoda, Dong-Qing Xu, Lei Qi and Masahiro Hirayama

Abstract

Neuroblastoma, originated from neural crest cells, is the most common extracranial solid tumor in childhood. Treatment is of limited utility for high-risk neuroblastoma and prognosis is poor. The high incidence of resistance of advanced-stage neuroblastoma to conventional therapies has prompt investigators to search for novel therapeutic approaches. Activation of IGF-R/PI3K/Akt/mTOR signaling pathway correlates with oncogenesis, poor prognosis, and chemotherapy resistance in neuroblastoma. Therefore, we investigated the effect of IGF-R/PI3K/Akt/mTOR signaling inhibitors in neuroblastoma. Significantly, IGF-R/PI3K/Akt/mTOR signaling inhibitors effectively inhibited cell growth and induced cell cycle arrest, autophagy, and apoptosis in neuroblastoma cells. Moreover, IGF-R/PI3K/Akt/mTOR signaling inhibitors significantly reduced tumor growth in mice xenograft model without apparent toxicity. Therefore, these results highlight the potential of IGF-R/PI3K/Akt/mTOR signaling pathway as a promising target for neuroblastoma treatment. Therefore, IGF-1R/PI3K/Akt/mTOR signaling inhibitors should be further investigated for treatment in clinical trials for high-risk neuroblastoma.

Keywords: neuroblastoma, insulin-like growth factor (IGF), phosphatidylinositol 3-kinase (PI3K), protein kinase-B (Akt), mammalian target of rapamycin (mTOR)

1. Introduction

Neuroblastoma (NB) is one of the most common extracranial solid tumors of early childhood [1, 2]. Prognosis of patients with NB depends on tumor stage, patient's age, and biologic feature of the tumor cells [3]. In patients under 1 year of age, NB is curable and sometimes spontaneously regress [4]. However, in older children with advanced stage, often the tumor is very aggressive, and patients have poor prognosis despite treatment with high-dose chemotherapy combined with autologous hematopoietic stem cell transplantation. Although immunotherapeutic therapy such as anti-GD2 monoclonal antibody has improved outcomes of advanced stage of NB, a number of patients still relapse and eventually die of disease [5]. The high incidence of resistance of high-risk NB to conventional therapies has prompted us to search for novel therapeutic approaches.

It was reported that phosphorylated protein kinase-B (Akt) correlates with poor patients' prognosis in NB [6], and the phosphatidylinositol-3-kinase (PI3K)/Akt/mammalian target of rapamycin (mTOR) pathway has subsequently been linked to augmented cell survival [7] and increased resistance to chemotherapy in NB [8].

IntechOpen

Therefore, targeting the PI3K/Akt/mTOR signaling pathway by appropriate inhibitors appears to be a promising strategy for overcoming therapy resistance [9].

Here, we demonstrate that NB cell lines are heterogeneous in their insulin growth factor-1 (IGF-1) receptor-mediated signaling [10]. The pattern of IGF-1 receptor/PI3K/Akt/mTOR pathway-mediated proliferation is an important determinant of the response to IGF-1 receptor antagonistic therapy in human NB [10]. Furthermore, our results highlight the potential of IGF-1 receptor/PI3K/Akt/mTOR signaling pathway as a promising target for NB treatment [10–13].

2. Heterogeneity of IGF-1 receptor/Akt pathway-mediated proliferation in NB

IGF-1, IGF-2, and insulin belong to a family of mitogenic growth factors and are involved in normal growth and differentiation of most tissues. The biological actions of both IGFs and insulin can be mediated by the IGF-1 receptor which is involved in mitogenic, anti-apoptotic, and oncogenic transforming responses [14, 15]. The IGF-1 receptor has two extracellular α-subunits and two intracellular β-subunits that form a heterotetrameric complex. Ligand interaction with α-subunits triggers the autophosphorylation of tyrosine kinase domains within the β-subunit [16–18]. The tyrosine kinase domains are connected to several intracellular pathways such as PI3K/Akt [19, 20]. Dysregulation of the IGF-1 receptor pathway is involved in promoting oncogenic transformation, cell proliferation, metastasis, angiogenesis, and resistance in numerous malignant diseases, such as multiple myeloma [21], carcinomas [22], and NB [23]. Since high cellular heterogeneity is a hallmark of NB, which may account for the wide range of clinical presentations and nonuniform response to treatment, we hypothesized that NB cells are heterogeneous in their IGF-1 receptor signaling-mediated cell proliferation. Thirty-one NB cell lines were cultured in three different conditions, insulin-containing serum-free medium (SFM), RPMI1640 without FBS (serum starvation medium), and RPMI1640 with 10% FBS (serum-containing medium). Based on the response patterns, 31 cell lines were subdivided into three groups [10]. Group 1, which consisted of three NB cell lines, could proliferate for more than 3 days in SFM, RPMI1640 without FBS, and RPMI1640 with 10% FBS [10]. Group 2, which consisted of 10 cell lines, could proliferate in SFM and RPMI1640 with 10% FBS but not in RPMI1640 [10]. Group 3, which consisted of 18 NB cell lines, proliferated only in RPMI1640 with 10% FBS [10]. NB cell proliferation in RPMI1640 in the presence of exogenous IGF (IGF-1, IGF-2) and insulin was examined. These IGF and insulin accelerated cell proliferation in group 1 and group 2 NB cell lines but not in group 3 NB cell lines [10]. Group 1 NB cell lines were able to proliferate in RPMI1640 in the absence of exogenous IGF and insulin [10].

3. Impairment of Akt activation and cell proliferation in NB by IGF-1 receptor inhibitor, picropodophyllin (PPP)

The IGF-1 receptor inhibitors, such as IGF-1 receptor-neutralizing antibodies and IGF-1 receptor antisense/siRNA, have been shown to block cancer cell proliferation [24]. Selective IGF-1 receptor inhibitor, picropodophyllin (PPP), lacks inhibitory activity on tyrosine phosphorylation of insulin receptor tyrosine kinase (RTK) [25]. Inhibition of the IGF-1 RTK with PPP is noncompetitive with respect to ATP, suggesting interference with the IGF-1 receptor at substrate level [26]. It is reported that PPP specifically blocks phosphorylation of the Tyr1136 residue in the activation loop of IGF-1 receptor kinase [27]. Inhibition of IGF-1 receptor with PPP has been demonstrated in a lot of cancers such as multiple myeloma [26], melanoma [28], breast cancer [29], and glioblastoma cells [30]. PPP inhibited Akt activation and suppressed cell proliferation in group 1 and group 2 NB cell lines but less in group 3 NB cell lines [10]. Elevation of ERK phosphorylation was only observed in group 1 NB cell lines [10]. U0126 is a MEK inhibitor effectively suppressed ERK phosphorylation, U0126 (2.5 μM) did not suppress cell proliferation induced by IGF-1, IGF-2, or insulin in group 1 and group 2 NB cell lines [10]. In groups 1 and 2 NB cell lines, IGF-1 receptor/PI3K/Akt pathway is critical for cell proliferation. Although IGF-1, IGF-2, and insulin activated Akt in group 3 NB cell line, cell proliferation was not increased in RPMI1640. This suggests that the activation of IGF-1R/Akt pathway is insufficient for cell proliferation [10]. Since IGF-1 and IGF-2 regulate apoptosis [31] and cell cycle progression [32], activation of caspase 3 and PARP was examined in NB cell lines. In group 1 NB cell lines, caspase 3 and PARP were not cleaved in RPMI1640 [10]. In group 2 NB cell lines, caspase 3 and PARP were cleaved in RPMI1640, and the cleavage of caspase 3 was suppressed by addition of IGF-1, IGF-2, and insulin [10]. In group 3 NB cell lines, cleavages of caspase 3 and PARP were observed in RPMI1640 [10]. However, they were not suppressed by IGF-1, IGF-2, and insulin, even though Akt activation was upregulated by IGF-1, IGF-2, and insulin [10]. Since PPP induced G2/M arrest and apoptosis by inhibiting IGF-1 receptor [26, 33], cell cycle phase distribution in NB cells was examined in the presence of PPP. PPP treatment (2.5 μM) for 12 hours increased the G2/M fraction and shifted cell cycle profile from G0-G1 dominant to G2/M dominant in group 1 and group 2 NB cell lines [10]. However, group 3 NB cell lines did not show G2/M arrest and the G0/G1 fraction was not affected [10]. Furthermore, cyclin B1, a marker protein of G2/M phase of cell cycle, was upregulated, and cyclin D1, a marker protein of G0/G1, synchronously declined in group 1 and group 2 NB cell lines [10]. However, accumulation of cyclin B1 and decline of cyclin D1 were not observed in group 3 NB cell lines [10]. This may be explained by the insensitiveness of group 3 cell lines to PPP-induced G2/M arrest.

4. Impairment of Akt activation and cell proliferation in NB by Akt inhibitor, MK2006

MK2206 selectively inhibits AKT and has potency against Akt1 and Akt2 isoforms than Akt3. In pediatric solid tumors, MK2206 is effective in vitro and in vivo [34, 35]. In clinical trials, stable disease was observed in different kinds of cancers [36, 37]. Our results also suggested that MK2206 (2.5 μM) completely inhibited Akt phosphorylation and cell proliferation in NB cell lines [10]. Furthermore, MK2206 also impaired the cell proliferation induced by exogenous IGF and insulin in NB cell lines [10].

5. Inhibition of NB cell proliferation by mTOR inhibitor

The PI3K/Akt/mTOR signaling cascade is one of the most important intracellular pathways, which is frequently activated in diverse cancers [38, 39]. The PI3K/Akt/mTOR pathway can be activated by transmembrane tyrosine kinase growth factor receptors, such as IGF-1 receptor, fibroblast growth factor receptors, ErbB family receptors, and others [40, 41]. Both mTOR S2448 and mTOR S2481 were extensively phosphorylated in NB cell lines [13]. Cell proliferation of NB cell lines was inhibited by AZD8055, a potent dual mTORC1-mTORC2 inhibitor [13]. According to the IC_{50}, the NB cell lines were divided into two groups, sensitive to AZD8055 group (IC_{50} < 0.5 μM) and insensitive to AZD8055 group (IC_{50} > 0.5 μM) [13]. We also found that insensitive group showed lower mTOR ($p < 0.001$) expression and lower activity of mTOR complex 1 ($p = 0.013$) and mTOR complex 2 ($p = 0.023$) [13]. Cell cycle distribution analysis of NB cell lines was performed by flow cytometry. Cell cycle was affected with an increase in G0/G1 phase in dose-dependent manner of AZD8055 [13]. Western blotting analysis revealed that Cyclin D1 and Cyclin D3 were downregulated in AZD8055-treated NB cells [13]. AZD8055 inhibited both mTOR S2448 and mTOR S2481 phosphorylation significantly in a concentration-dependent manner [13]. Phosphorylation of downstream targets of mTOR complex 1, P70S6K T389 and 4E-BP1 S65, was also inhibited by AZD8055 treatment [13]. AZD8055 inhibited mTOR complex 2 substrates, Akt S473 and Akt T308 [13]. Although phosphorylation of Akt S473 was persistently inhibited in response to AZD8055 treatment, phosphorylation of Akt at the T308 site was inhibited for only 3–6 hours. These results indicate that AZD8055 inhibits mTOR activity and its downstream proteins in vitro in NB cells. Interestingly, NB cell lines were induced autophagy by AZD8055 treatment via downregulation of Akt/mTOR signaling pathway [13]. Autophagy inhibitor, 3-methyladenine, treatment resulted in a significant decrease of the AZD8055-induced apoptosis [13]. These results suggest that AZD8055 inhibited cell growth and induced cell cycle arrest, autophagy, and apoptosis in NB cells. Moreover, NB tumor growth in athymic nude mice was significantly inhibited by AZD8055 without toxicity [13]. Taken together, our results highlight that mTOR is a promising target for NB treatment. AZD8055 might be investigated for treatment of patients with advanced and refractory NB.

6. AZD8055 treatment in combination with MEK/ERK inhibitor

The problem that will plague single-target drugs is the cancer's ability to activate alternative survival pathways leading to drug resistance and toxicity even in the multimodality setting. In addition, activation of multiple signaling pathways effectively causes cancer cell proliferation and survival. For example, RAS and PI3KCA are concurrently activated in melanoma, lung, and colorectal cancers [42–44]. These results suggested that combination therapies targeted on multiple signaling pathways could be more effective than targeting either pathway alone. Although AZD8055 is a promising drug in NB treatment [13], AZD8055-resistant NB sublines were acquired by prolonged stepwise exposure. After incubation with AZD8055 for 4–12 weeks, two acquired AZD8055-resistant sublines proliferated stably in RPMI1640 plus 10% FBS medium in the presence of AZD8055 (3 μM) [12]. The AZD8055-resistant sublines exhibited marked resistance to AZD8055 compared to the parent cells, and IC_{50} of the resistant sublines was 60–100 times higher than the parent NB cell lines [12]. By cell cycle analysis, accumulation of S phase was observed, and cyclin D3 and CDK4 were upregulated in AZD8055-resistant sublines [12]. Although AZD8055 treatment inhibited MEK/ERK activation in parent cells, MEK/ERK phosphorylation was continued despite AZD8055 treatment in resistant cells [12]. The combination therapy of AZD8055 and MEK/ERK inhibitor U0126 significantly inhibited cell proliferation compared to U0126 monotherapy [12], suggesting that combination therapy can overcome AZD8055 resistance. Furthermore, in athymic mice model, AZD8055 and U0126 co-treatment was more efficient to suppress resistant NB tumor growth compared to U0126 monotherapy [12].

7. MK2206 treatment in combination with PDK1 inhibitor

MK2206 treatment induced a dose-dependent inhibition of cell proliferation, with IC_{50} ranging from 1.22 to 4.35 μM in NB cell lines [11]. MK2206-resistant cells were induced by stepwise escalation of MK-2206 exposure (4–12 weeks) [11]. These cells proliferated in RPMI1640 plus 10% FBS medium in the presence of MK2206 (5 μM), while cell death was induced in parent cells [11]. IC_{50} of the resistant cell lines was 6–7 times higher than the parent NB cell lines [11]. Small-molecule GSK2334470 selectively inhibits 3-phosphoinositide-dependent protein kinase 1 (PDK1) with low concentration but does not suppress the activity of other protein kinases at higher concentrations [45]. Although GSK2334470 attenuated cell proliferation in both parent cells and MK2206-resistant sublines, IC_{50} of GSK2334470 in resistant sublines was lower than that of parent cell lines [11]. GSK2334470 induced G0-G1 accumulation of cell cycle phase distribution in parent cell lines [11]. In MK2206-resistant sublines, G0-G1 accumulation induced by GSK2334470 was higher than parent cell lines [11].

8. Effect of NB MYCN status on susceptibility to AZD8055

Amplification of the MYCN oncogene is the most powerful single predictor of adverse outcome of NB [46]. MYCN amplification is observed in about 20% of all NB patients and is usually associated with fatal outcome of the disease [47]. Schramm et al. demonstrated transcriptomal upregulation of mTOR-related genes by MYCN [48]. MYCN-driven NB in mice displayed activation of the mTOR pathway on the protein level, and activation of MYCN in NB cells resulted in high sensitivity toward mTOR inhibition [48]. Therefore, it is examined whether MYCN

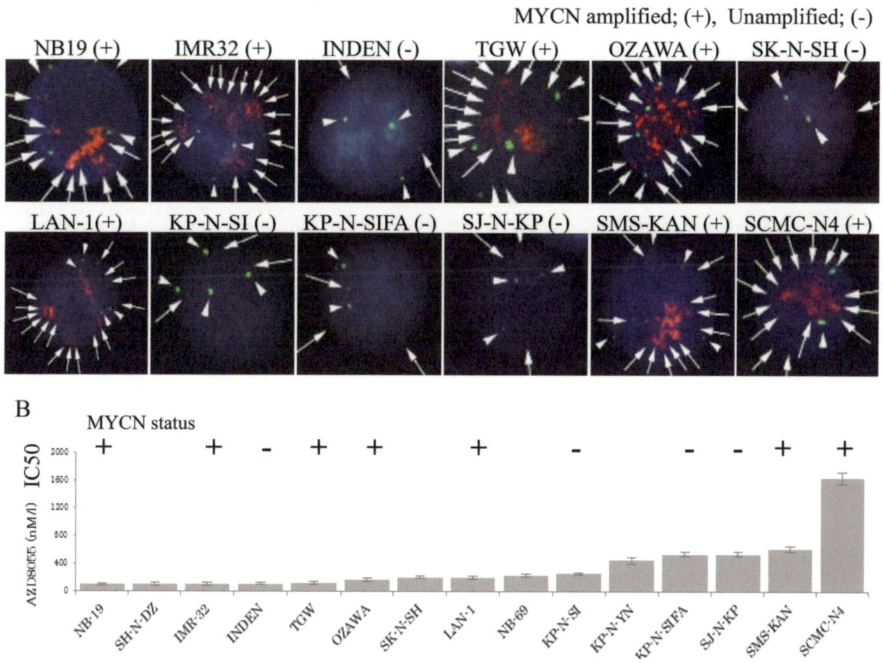

Figure 1.
Effect of MYCN status on susceptibility to AZD8055. (A) MYCN gene in tested 12 NB cell lines. MYCN status was defined as previously [53]. No alteration: cells with two MYCN signals and two CEP2 signals; amplification: the number of MYCN signals is at least 10 copies greater than the control probe signals; loss/imbalance: presence of at least two MYCN signals and increased CEP2 signals; gain: the number of MYCN signals is 1–9 copies more than CEP2 signals. (B) Fifteen NB cell lines were treated with AZD8055 at different concentrations in RPMI1640 + 10% FBS. Cell growth was evaluated as cell numbers at 72 hours. Data are expressed as the mean ± SD. IC_{50} (half maximal inhibitory concentration) of the 15 NB cell lines was calculated depending on the MTT results.

status of NB cell lines affects susceptibility to AZD8055. Using fluorescence in situ hybridization (FISH), we observed MYCN amplification in all investigated nuclei of NB19, IMR32, TGW, OZAWA, LAN-1, SMS-KAN, and SCMC-N4 cell lines (**Figure 1A**). As shown in **Figure 1A**, MYCN amplification was not observed in INDEN (loss/imbalance), SK-N-SH (gain), KP-N-SI (loss/imbalance), KP-N-SIFA (gain), and SJ-N-KP (loss/imbalance). As shown in **Figure 1B**, the relationship between MYCN status and susceptibility to AZD8055 was not found.

9. ALK as a therapeutic target in NB

Constitutive activation of the ALK receptor tyrosine kinase by mutation or translocation appears to contribute to the malignant phenotype of several cancers, including NB, making it a potentially therapeutic target [49]. Both wild type ALK and the F1174L-mutated ALK upregulate MYCN expression [50], and these two genes lead to constitutive phosphorylation of ALK and of downstream signaling molecules, PI3K/AKT/mTOR, that are critical for cell proliferation and survival [51]. ALK^{F1174L} cosegregates with MYCN amplification in patients, and this combination is associated with a particularly poor prognosis, as shown by the fatal outcome of 9 of 10 children with ALK^{F1174L}/MYCN-amplified tumors [52]. Targeting ALK with PI3K/Akt/mTOR inhibitors may inhibit cell growth in tumor lines with concomitant MYCN amplification.

10. Conclusions

NB cell lines can be categorized into three groups by the patterns of IGF-1R/Akt pathway response. PPP, MK2206, and AZD8055 showed significant antitumor effect in NB cells. Our current results highlight the potential of IGF-1R/PI3K/Akt/mTOR pathway as a promising target for NB treatment. PPP, MK2206, and AZD8055 should be further investigated for NB treatment in clinical trials.

Conflict of interest

The authors declare no potential conflicts of interest.

Author details

Hidemi Toyoda*, Dong-Qing Xu, Lei Qi and Masahiro Hirayama
Department of Pediatrics, Mie University Graduate School of Medicine,
Tsu, Mie, Japan

*Address all correspondence to: htoyoda@clin.medic.mie-u.ac.jp

References

[1] Louis CU, Shohet JM. Neuroblastoma: Molecular pathogenesis and therapy. Annual Review of Medicine. 2015;**66**: 49-63

[2] Maris JM. Recent advances in neuroblastoma. The New England Journal of Medicine. 2010;**362**(23):2202-2211

[3] Erttmann R, Tafese T, Berthold F, Kerbl R, Mann J, Parker L, et al. 10 years' neuroblastoma screening in Europe: Preliminary results of a clinical and biological review from the Study Group for Evaluation of Neuroblastoma Screening in Europe (SENSE). European Journal of Cancer. 1998;**34**(9):1391-1397

[4] D'Angio GJ, Evans AE, Koop CE. Special pattern of widespread neuroblastoma with a favourable prognosis. Lancet. 1971;**1**(7708):1046-1049

[5] Yu AL, Gilman AL, Ozkaynak MF, London WB, Kreissman SG, Chen HX, et al. Anti-GD2 antibody with GM-CSF, interleukin-2, and isotretinoin for neuroblastoma. The New England Journal of Medicine. 2010;**363**(14):1324-1334

[6] Opel D, Poremba C, Simon T, Debatin KM, Fulda S. Activation of Akt predicts poor outcome in neuroblastoma. Cancer Research. 2007;**67**(2):735-745

[7] Boller D, Schramm A, Doepfner KT, Shalaby T, von Bueren AO, Eggert A, et al. Targeting the phosphoinositide 3-kinase isoform p110 delta impairs growth and survival in neuroblastoma cells. Clinical Cancer Research. 2008;**14**(4):1172-1181

[8] Li Z, Thiele CJ. Targeting Akt to increase the sensitivity of neuroblastoma to chemotherapy: Lessons learned from the brain-derived neurotrophic factor/TrkB signal transduction pathway. Expert Opinion on Therapeutic Targets. 2007;**11**(12):1611-1621

[9] Paz-Ares L, Blanco-Aparicio C, Garcia-Carbonero R, Carnero A. Inhibiting PI3K as a therapeutic strategy against cancer. Clinical & Translational Oncology. 2009;**11**(9): 572-579

[10] Qi L, Toyoda H, Shankar V, Sakurai N, Amano K, Kihira K, et al. Heterogeneity of neuroblastoma cell lines in insulin-like growth factor 1 receptor/Akt pathway-mediated cell proliferative responses. Cancer Science. 2013;**104**(9):1162-1171

[11] Qi L, Toyoda H, Xu DQ, Zhou Y, Sakurai N, Amano K, et al. PDK1-mTOR signaling pathway inhibitors reduce cell proliferation in MK2206 resistant neuroblastoma cells. Cancer Cell International. 2015;**15**:91

[12] Xu DQ, Toyoda H, Qi L, Morimoto M, Hanaki R, Iwamoto S, et al. Induction of MEK/ERK activity by AZD8055 confers acquired resistance in neuroblastoma. Biochemical and Biophysical Research Communications. 2018;**499**(3):425-432

[13] Xu DQ, Toyoda H, Yuan XJ, Qi L, Chelakkot VS, Morimoto M, et al. Anti-tumor effect of AZD8055 against neuroblastoma cells in vitro and in vivo. Experimental Cell Research. 2018;**365**(2):177-184

[14] Bentov I, Werner H. IGF, IGF receptor and overgrowth syndromes. Pediatric Endocrinology Reviews. 2004;**1**(4):352-360

[15] Werner H, Bruchim I. The insulin-like growth factor-I receptor as an oncogene. Archives of Physiology and Biochemistry. 2009;**115**(2):58-71

[16] Brodt P, Samani A, Navab R. Inhibition of the type I insulin-like growth factor receptor expression and signaling: Novel strategies for antimetastatic therapy. Biochemical Pharmacology. 2000;**60**(8):1101-1107

[17] Gronborg M, Wulff BS, Rasmussen JS, Kjeldsen T, Gammeltoft S. Structure-function relationship of the insulin-like growth factor-I receptor tyrosine kinase. The Journal of Biological Chemistry. 1993;**268**(31):23435-23440

[18] Laviola L, Natalicchio A, Giorgino F. The IGF-I signaling pathway. Current Pharmaceutical Design. 2007;**13**(7):663-669

[19] Pollak M. Insulin and insulin-like growth factor signalling in neoplasia. Nature Reviews Cancer. 2008;**8**(12):915-928

[20] Sartelet H, Oligny LL, Vassal G. AKT pathway in neuroblastoma and its therapeutic implication. Expert Review of Anticancer Therapy. 2008;**8**(5):757-769

[21] Mahindra A, Cirstea D, Raje N. Novel therapeutic targets for multiple myeloma. Future Oncology. 2010;**6**(3):407-418

[22] Barlaskar FM, Spalding AC, Heaton JH, Kuick R, Kim AC, Thomas DG, et al. Preclinical targeting of the type I insulin-like growth factor receptor in adrenocortical carcinoma. The Journal of Clinical Endocrinology and Metabolism. 2009;**94**(1):204-212

[23] Zumkeller W, Schwab M. Insulin-like growth factor system in neuroblastoma tumorigenesis and apoptosis: Potential diagnostic and therapeutic perspectives. Hormone and Metabolic Research. 1999;**31**(2-3):138-141

[24] Surmacz E. Growth factor receptors as therapeutic targets: Strategies to inhibit the insulin-like growth factor I receptor. Oncogene. 2003;**22**(42):6589-6597

[25] Girnita A, Girnita L, del Prete F, Bartolazzi A, Larsson O, Axelson M. Cyclolignans as inhibitors of the insulin-like growth factor-1 receptor and malignant cell growth. Cancer Research. 2004;**64**(1):236-242

[26] Stromberg T, Ekman S, Girnita L, Dimberg LY, Larsson O, Axelson M, et al. IGF-1 receptor tyrosine kinase inhibition by the cyclolignan PPP induces G2/M-phase accumulation and apoptosis in multiple myeloma cells. Blood. 2006;**107**(2):669-678

[27] Vasilcanu D, Girnita A, Girnita L, Vasilcanu R, Axelson M, Larsson O. The cyclolignan PPP induces activation loop-specific inhibition of tyrosine phosphorylation of the insulin-like growth factor-1 receptor. Link to the phosphatidyl inositol-3 kinase/Akt apoptotic pathway. Oncogene. 2004;**23**(47):7854-7862

[28] Karasic TB, Hei TK, Ivanov VN. Disruption of IGF-1R signaling increases TRAIL-induced apoptosis: A new potential therapy for the treatment of melanoma. Experimental Cell Research. 2010;**316**(12):1994-2007

[29] Klinakis A, Szabolcs M, Chen G, Xuan S, Hibshoosh H, Efstratiadis A. Igf1r as a therapeutic target in a mouse model of basal-like breast cancer. Proceedings of the National Academy of Sciences of the United States of America. 2009;**106**(7):2359-2364

[30] Yin S, Girnita A, Stromberg T, Khan Z, Andersson S, Zheng H, et al. Targeting the insulin-like growth factor-1 receptor by picropodophyllin as a treatment option for glioblastoma. Neuro-Oncology. 2010;**12**(1):19-27

[31] Varela-Nieto I, Hartl M, Gorospe I, Leon Y. Anti-apoptotic actions of

insulin-like growth factors: Lessons from development and implications in neoplastic cell transformation. Current Pharmaceutical Design. 2007;**13**(7):687-703

[32] Dupont J, Pierre A, Froment P, Moreau C. The insulin-like growth factor axis in cell cycle progression. Hormone and Metabolic Research. 2003;**35**(11-12):740-750

[33] Tazzari PL, Tabellini G, Bortul R, Papa V, Evangelisti C, Grafone T, et al. The insulin-like growth factor-I receptor kinase inhibitor NVP-AEW541 induces apoptosis in acute myeloid leukemia cells exhibiting autocrine insulin-like growth factor-I secretion. Leukemia. 2007;**21**(5):886-896

[34] Gorlick R, Maris JM, Houghton PJ, Lock R, Carol H, Kurmasheva RT, et al. Testing of the Akt/PKB inhibitor MK-2206 by the pediatric preclinical testing program. Pediatric Blood & Cancer. 2012;**59**(3):518-524

[35] Li Z, Yan S, Attayan N, Ramalingam S, Thiele CJ. Combination of an allosteric Akt Inhibitor MK-2206 with etoposide or rapamycin enhances the antitumor growth effect in neuroblastoma. Clinical Cancer Research. 2012;**18**(13):3603-3615

[36] Fouladi M, Perentesis JP, Phillips CL, Leary S, Reid JM, McGovern RM, et al. A phase I trial of MK-2206 in children with refractory malignancies: A children's oncology group study. Pediatric Blood & Cancer. 2014;**61**(7):1246-1251

[37] Yap TA, Yan L, Patnaik A, Fearen I, Olmos D, Papadopoulos K, et al. First-in-man clinical trial of the oral pan-AKT inhibitor MK-2206 in patients with advanced solid tumors. Journal of Clinical Oncology. 2011;**29**(35):4688-4695

[38] Liu P, Cheng H, Roberts TM, Zhao JJ. Targeting the phosphoinositide 3-kinase pathway in cancer. Nature Reviews Drug Discovery. 2009;**8**(8):627-644

[39] Polivka J Jr, Janku F. Molecular targets for cancer therapy in the PI3K/AKT/mTOR pathway. Pharmacology & Therapeutics. 2014;**142**(2):164-175

[40] Knuefermann C, Lu Y, Liu B, Jin W, Liang K, Wu L, et al. HER2/PI-3K/Akt activation leads to a multidrug resistance in human breast adenocarcinoma cells. Oncogene. 2003;**22**(21):3205-3212

[41] Stern DF. ERBB3/HER3 and ERBB2/HER2 duet in mammary development and breast cancer. Journal of Mammary Gland Biology and Neoplasia. 2008;**13**(2):215-223

[42] Hoeflich KP, Merchant M, Orr C, Chan J, Den Otter D, Berry L, et al. Intermittent administration of MEK inhibitor GDC-0973 plus PI3K inhibitor GDC-0941 triggers robust apoptosis and tumor growth inhibition. Cancer Research. 2012;**72**(1):210-219

[43] Smalley KS, Haass NK, Brafford PA, Lioni M, Flaherty KT, Herlyn M. Multiple signaling pathways must be targeted to overcome drug resistance in cell lines derived from melanoma metastases. Molecular Cancer Therapeutics. 2006;**5**(5):1136-1144

[44] Sos ML, Fischer S, Ullrich R, Peifer M, Heuckmann JM, Koker M, et al. Identifying genotype-dependent efficacy of single and combined PI3K- and MAPK-pathway inhibition in cancer. Proceedings of the National Academy of Sciences of the United States of America. 2009;**106**(43):18351-18356

[45] Najafov A, Sommer EM, Axten JM, Deyoung MP, Alessi DR. Characterization of GSK2334470, a novel and highly specific inhibitor of PDK1. The Biochemical Journal. 2011;**433**(2):357-369

[46] Goto S, Umehara S, Gerbing RB, Stram DO, Brodeur GM, Seeger RC, et al. Histopathology (International Neuroblastoma Pathology Classification) and MYCN status in patients with peripheral neuroblastic tumors: A report from the children's cancer group. Cancer. 2001;**92**(10):2699-2708

[47] Seeger RC, Brodeur GM, Sather H, Dalton A, Siegel SE, Wong KY, et al. Association of multiple copies of the N-myc oncogene with rapid progression of neuroblastomas. The New England Journal of Medicine. 1985;**313**(18):1111-1116

[48] Schramm A, Koster J, Marschall T, Martin M, Schwermer M, Fielitz K, et al. Next-generation RNA sequencing reveals differential expression of MYCN target genes and suggests the mTOR pathway as a promising therapy target in MYCN-amplified neuroblastoma. International Journal of Cancer. 2013;**132**(3):E106-E115

[49] Davidoff AM. Neuroblastoma. Seminars in Pediatric Surgery. 2012;**21**(1):2-14

[50] Reiff T, Huber L, Kramer M, Delattre O, Janoueix-Lerosey I, Rohrer H. Midkine and Alk signaling in sympathetic neuron proliferation and neuroblastoma predisposition. Development (Cambridge, England). 2011;**138**(21):4699-4708

[51] Matthay KK, George RE, Yu AL. Promising therapeutic targets in neuroblastoma. Clinical Cancer Research. 2012;**18**(10):2740-2753

[52] De Brouwer S, De Preter K, Kumps C, Zabrocki P, Porcu M, Westerhout EM, et al. Meta-analysis of neuroblastomas reveals a skewed ALK mutation spectrum in tumors with MYCN amplification. Clinical Cancer Research. 2010;**16**(17):4353-4362

[53] Ambros PF, Ambros IM. Pathology and biology guidelines for resectable and unresectable neuroblastic tumors and bone marrow examination guidelines. Medical and Pediatric Oncology. 2001;**37**(6):492-504

Chapter 6

Anticancer Drugs' Deoxyribonucleic Acid (DNA) Interactions

Saad Hmoud Alotaibi and Awad Abdalla Momen

Abstract

The deoxyribonucleic acid (DNA) is the molecule of life that controls all the chemical changes that take place in cells. The interaction of drugs with DNA is among the most important aspects of biological studies in drug discovery and pharmaceutical development processes. Moreover, the knowledge of specific targets in rational design of chemotherapeutics is a fundamental factor, principally, for the design of molecules that can be used in the treatment of oncologic diseases. Observing the pre- and postsigns of drug-DNA interaction provides good evidence for the interaction mechanism to be elucidated. Also, this interaction could be used for the quantification of drugs and for the determination of new drugs targeting DNA. Approaches can provide new insight into rational drug design and would lead to further understanding of the interaction mechanism between anticancer drugs and DNA. The intention of this chapter is to provide several examples of anticancer drugs, DNA interaction, and the mechanisms of interaction in order to understand the influence of several interaction factors in the capacity and selectivity of the anticancer drugs to interact with DNA. In addition, different experimental and theoretical approaches to detect and to evaluate the anticancer drugs' interactions with DNA were also discussed.

Keywords: anticancer drugs, DNA, interactions, proteins, hydrogen bond

1. Introduction

DNA information is stored in the form of a code that constitutes four chemical bases namely: cytosine (C), adenine (A), thymine (T), and, lastly, guanine (G). The human DNA has approximately 3 billion bases, and not <99% of these bases are similar in all individuals. The sequence of these bases governs the available information for maintaining and building an organism, similar to the manner in which alphabetical letters are arranged to form sentences and words [1].

The chemical bases in a DNA pair up (C with G and A with T), in order to produce units known as base pairs. In addition, each base is attached to a phosphate molecule and a sugar molecule. Together, a phosphate, sugar, and base are referred to as a nucleotide. The nucleotides are organized in two long strands thereby forming a spiral known as a double helix. A double helix's structure resembles a ladder, with the phosphate and sugar molecules forming the ladder's vertical sidepieces. On the other hand, the base pairs form the rungs of the ladder.

Many anticancer drugs in clinical use interact with DNA through intercalation, which can be defined as the process by which compounds containing planar aromatic or heteroaromatic ring systems are inserted between adjacent base pairs perpendicularly to the axis of the helix and without disturbing the overall stacking pattern due to Watson-Crick hydrogen bonding [2, 3].

2. Structural features of DNA

DNA consists of two complementary anti-parallel sugar phosphate poly-deoxyribonucleotide strands that are associated with specific hydrogen bonding between nucleotide bases. The two strands are held together primarily through Watson-Crick hydrogen bonds where A forms two hydrogen bonds with T and C forms three hydrogen bonds with G (**Figure 1**). The structure of these paired strands defines the helical grooves, within which the edges of the heterocyclic bases are exposed. The biologically relevant B-form of the DNA double helix is characterized by a shallow-wide major groove and a deep-narrow minor groove. The chemical structure (feature) of the molecular surfaces in a given DNA sequence is well known in either groove. This forms the basis for molecular recognition of duplex DNA by small molecules and proteins [4, 5].

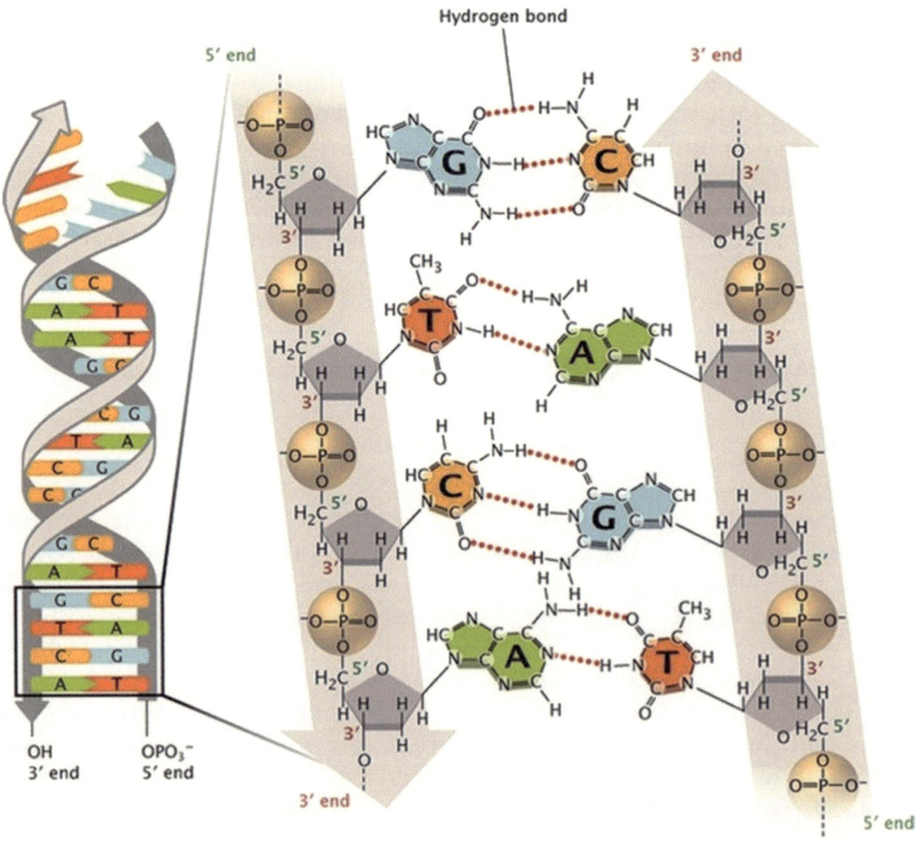

Figure 1.
Watson-Crick pairing between purine and pyrimidine bases in complementary DNA strand.

3. Anticancer drug-DNA interaction

DNA as carrier of genetic information is a major target for anticancer drug interaction because of the ability to interfere with transcription and DNA replication, a major step in cell growth and division. There are three principally different ways of anticancer drug binding. First is through control of transcription factors and polymerases. Here, the anticancer drugs interact with the proteins that bind directly to DNA. Second is through RNA binding to DNA double helices to form nucleic acid triple helical structures or RNA hybridization to exposed DNA single strand regions that will be forming DNA-RNA hybrids and it may interfere with transcriptional activity. Third is through small aromatic ligand molecules that bind to DNA double helical structures through non-covalent interaction either by intercalating binder or by minor groove binders (**Figure 2**) [6, 7]. Therefore, intercalation can be defined as the process by which compounds containing planar aromatic or heteroaromatic ring systems are inserted between adjacent base pairs perpendicularly to the axis of the helix and without disturbing the overall stacking pattern due to Watson-Crick hydrogen bonding [8]. In addition, intercalation binding involves the insertion of a planar molecule between DNA base pairs, which results in a decrease in the DNA helical twist and lengthening of the DNA. While groove binding, unlike intercalation, does not induce large conformational changes in DNA and may be considered similar to standard lock-and-key models for ligand-macromolecular binding. In addition, Groove binders are usually crescent-shaped molecules that bind to the minor groove of DNA [7].

In order to accommodate the binder (like intercalation binder), DNA must undergo a conformational change to create a cavity for the incoming chromophore. The double helix is therefore partially unwound, which leads to distortions of the sugar-phosphate backbone and changes in the twist angle between successive base pairs (**Figure 3**) [8]. Once the drug has been sandwiched between the DNA base

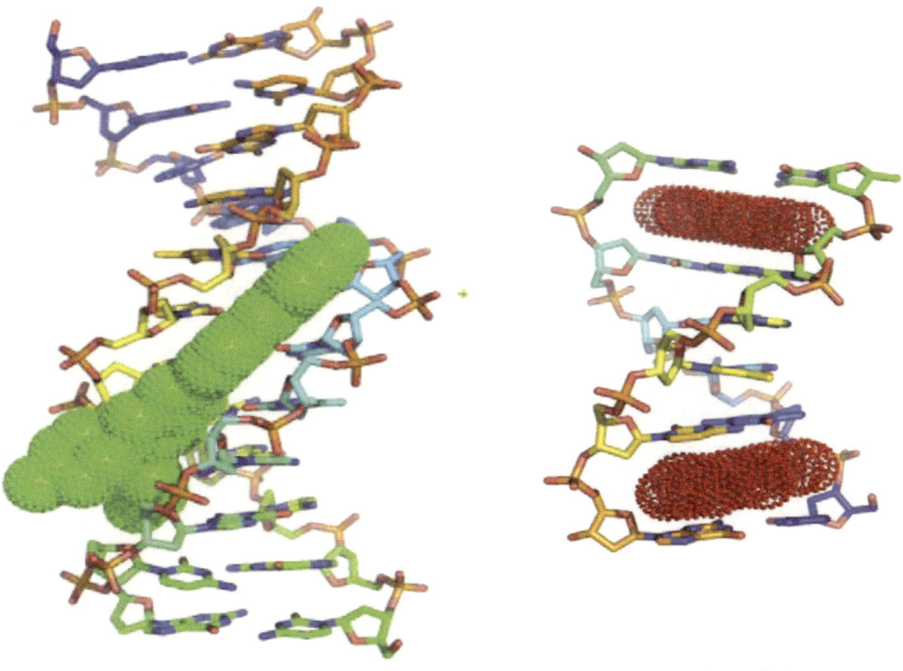

Figure 2.
Groove binding to the minor groove of DNA (left) and the intercalation into DNA (right).

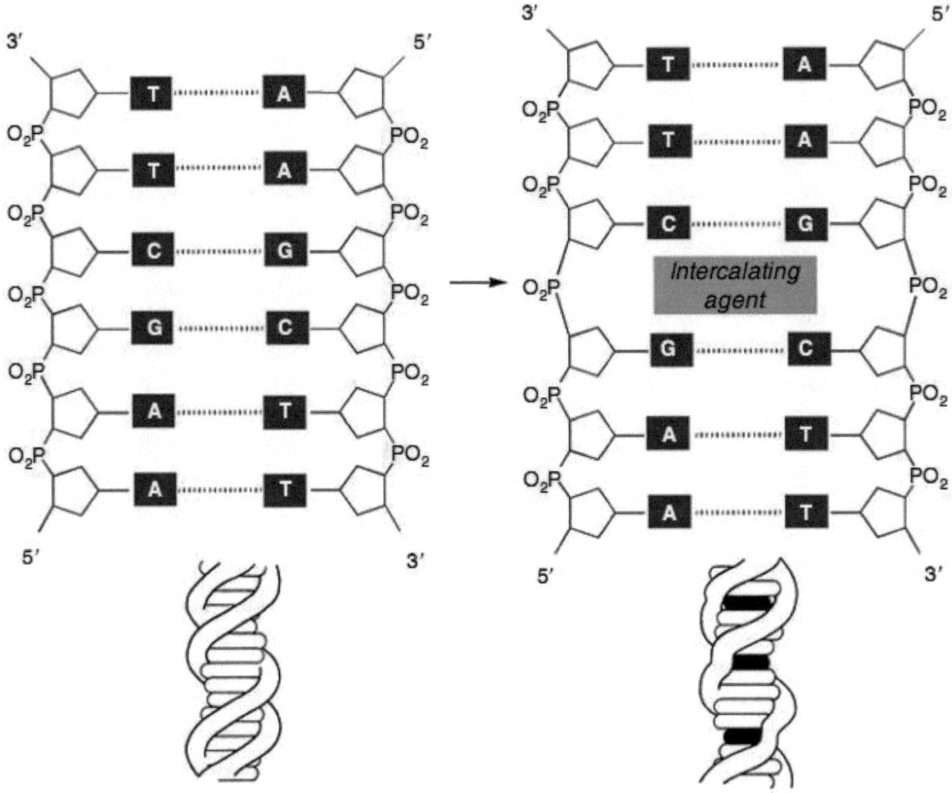

Figure 3.
Deformation of DNA by an intercalating agent.

pairs, several non-covalent interactions such as Van der Waals interaction and hydrogen bonding optimizes the stability of the complex.

4. Types of drug-DNA interactions

The study of interaction between drug molecules and DNA is very exciting and significant not only in understanding the mechanism of interaction, but also for the design of new drugs. However, the mechanism of interactions between them is still relatively little known. By understanding the mechanism of interaction between them, designing of new DNA-targeted drugs and the screening of these in vitro will be possible [9]. Many of the most valuable anticancer drugs currently used in therapy interact with DNA either by a covalent or non-covalent mechanism. Unfortunately, several of them show a considerable toxicity when the DNA molecular target is present in both normal and tumor cells [10]. The covalent type of binding of drug-DNA is irreversible and invariably causes the complete inhibition of DNA processes and subsequent cell death. A major advantage of covalent binders is the high binding strength. However, covalent bulky adducts can cause DNA backbone distortion, which affect both transcription and replication (disrupting protein complex recruitment). The covalent binders are also called alkylating agents due to adduct formation because they are used in cancer treatment to attach an alkyl group (C_nH_{2n+1}) to DNA [11]. **Table 1** lists the different types of drug-DNA interactions with suitable examples. In addition, some important examples of a cross-linking agent covalent and non-covalent binder were shown in **Figure 4** [5, 12].

No.	Type of interaction	Example
1	Covalent bonding	Nitrogen mustard, carboplatin and cyclophosphamide
2	Non-covalent bonding	Ethidium bromide and quinacrine

Table 1.
Listing the different types of drug-DNA interactions with suitable examples.

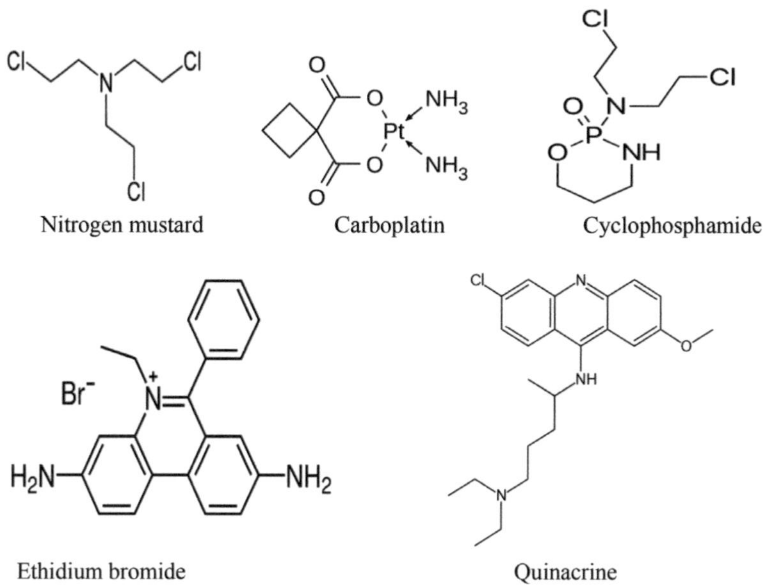

Figure 4.
Chemical structure of some covalent and non-covalent binders of DNA.

Non-covalent DNA interacting agents (groove, intercalators, and external binders) are generally considered less cytotoxic than agents producing covalent DNA adducts and other DNA damage. The non-covalent binding type is reversible and is typically preferred over covalent adduct formation keeping the drug metabolism and toxic side effects in mind. In addition, non-covalent DNA interacting agents can changes DNA conformation, DNA torsional tension, interrupt protein-DNA interaction, and potentially lead to DNA strand breaks [11].

5. Modeling of hairpin minor groove binders

Hairpin minor grove binding molecules have been identified and synthesized that bind to G-C reach nucleotide sequences. Hairpin polyamides are linked systems that exploit a set of simple recognition rules for DNA base pairs through specific orientation of imidazole (Im) and pyrrole (Py) rings (**Figure 5**) [13]. They originated from the discovery of the three-ring Im-Py-Py molecule that bound to minor groove DNA as an antiparallel side by side dimer.

The solid phase synthesis of polyamides of variable length has produced efficient ligands. The advantage of polyamide ligand design has been reached with finding structures able to recognize DNA sequences of specific genes. Moreover, a new strategy of rational drug design exploits the combination of polyamides with bis-intercalating structures. The new synthetic compound showed a resistant against multidrug resistance in which small aromatic compounds are efficiently

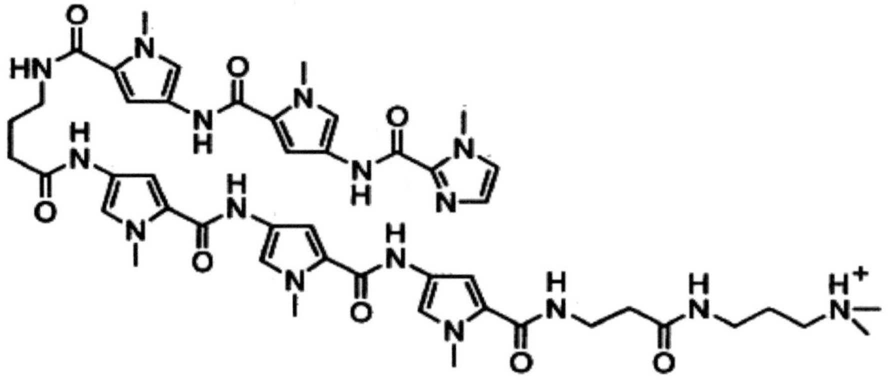

Figure 5.
Structure of hairpin polyamide Im-Py-Py.

expelled from the cell-by-cell membrane transport proteins that commonly referred to as ABC transporters or ATP binding cassette proteins [14].

6. Rational for drug design

When a compound intercalates into nucleic acids, there are changes, which occur in both the DNA and the compound during complex formation that can be used to study the ligand DNA interaction. The binding is of course an equilibrium process because no covalent bond formation is involved. The binding constant can be determined by measuring the free and DNA bound form of the ligand. In addition, DNA double helix structures are found to be more stable with intercalating agents present and show a reduced heat denaturation. Correlating these biophysical parameters with cytotoxicity is used to support the antitumor activity of these drugs as based on their ability to intercalate in DNA double helical structures. [15].

Improvement of anticancer drugs based on intercalating activity is not only focused on DNA-ligand interaction, but also on tissue distribution and toxic side effects on the heart (cardiac toxicity) due to redox reduction of the aromatic rings and subsequent free radical formation. Free radical species are thought to induce destructive cellular events such as enzyme inactivation, DNA strand cleavage and membrane lipid peroxidation [16, 17].

7. Cisplatin-DNA interactions

Cisplatin (cis-$[PtCl_2(NH_3)_2]$) is the most widely used anticancer drug today. Since the development of cisplatin became one of the main biological targets for the antitumor compounds. It is used against ovarian, cervical, head and neck, esophageal and non-small cell lung cancer. However, chemotherapy treatment by cisplatin comes with a price of severe side effects including nausea, vomiting and ear damage, as cisplatin not only attacks cancer cells, but also healthy cells. It is therefore important to elucidate the details of the cisplatin mode of action to design new cisplatin analogs that specifically target cancer cells. Furthermore, most cancer cells are insensitive towards cisplatin or develop resistance. There is therefore, also a need for cisplatin analogues with a broader range of cytotoxicity. The search for new analogues and the elucidation of the complete mode of action have been going on for more than 40 years and there is an enormous amount of data available for researchers. Still, the picture of how cisplatin works is incomplete [11, 18].

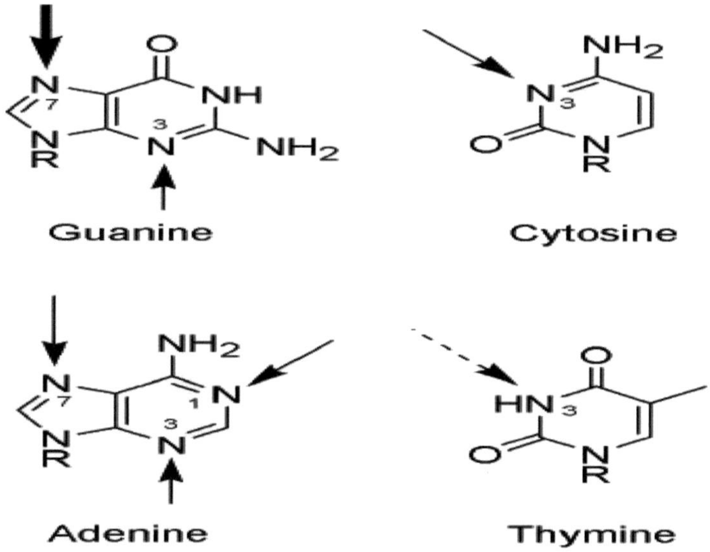

Figure 6.
The structure of the most common binding sites on the nucleobases for Pt. The big arrow on guanine indicates the overall favorable coordination site in DNA, the arrow towards thymine is dotted because the proton has to be removed before Pt association.

Cellular DNA has been shown to be the primary target for cisplatin, although cisplatin can react with several other cellular components. In the cell, the salt concentration is significantly lower (~20 mM) and cis-$[PtCl_2(NH_3)_2]$ is hydrolyzed by high salt concentration (>100 mM) to the probable active species cis-$[PtCl(OH_2)(NH_3)_2]^+$. The hydrolyzed product binds to DNA and preferentially to guanine N7> > adenine N7 > cytosine N3, first as a monoadduct, then forming a bidentate adduct. The primary products are 1,2-intrastrand cross-links of GpG (60–65%) or ApG (20–25%) sequences. A smaller amount corresponds to 1,3-intrastrand or G N7–G N7 interstrand adducts. The most common binding sites on the nucleobases for Pt are shown in **Figure 6** [18]. The big arrow on guanine indicates the overall favorable coordination site in DNA, the arrow towards thymine is dotted because the proton has to be removed before Pt association.

The formation of these 1,2-intrastrand cross-links alters the duplex conformation. The most dramatic effect is unwinding of the two strands and bending of the DNA double helix (several values for the bend angle are reported in the range 20–80°). The platinated adducts are assumed to be recognized by proteins, followed either by stabilization of the distorted DNA structure or removal of the lesion through repair [18]. The deformation of the DNA structure can interfere with the normal functions of DNA, such as replication and transcription, leading to cellular death by apoptosis or necrosis. The ineffective isomer of cisplatin, transplatin (trans-$[PtCl_2(NH_3)_2]$), is not able to form 1,2-intrastrand cross-links [19]. Transplatin forms only 1,3-intrastrand and interstrand cross-links and this might be the reason why transplatin is antitumor inactive [18].

In addition, a sensing system based on the photoinduced electron transfer of quantum dots (QDs) was also designed to measure the interaction of anticancer drug and DNA, taking mitoxantrone (MTX) as a model drug. The MTX adsorbed on the surface of QDs and this, can quench the photoluminescence (PL) of QDs through the photoinduced electron-transfer process, then the addition of DNA will bring the restoration of QDs PL intensity, as DNA can bind with MTX and remove it from QDs.

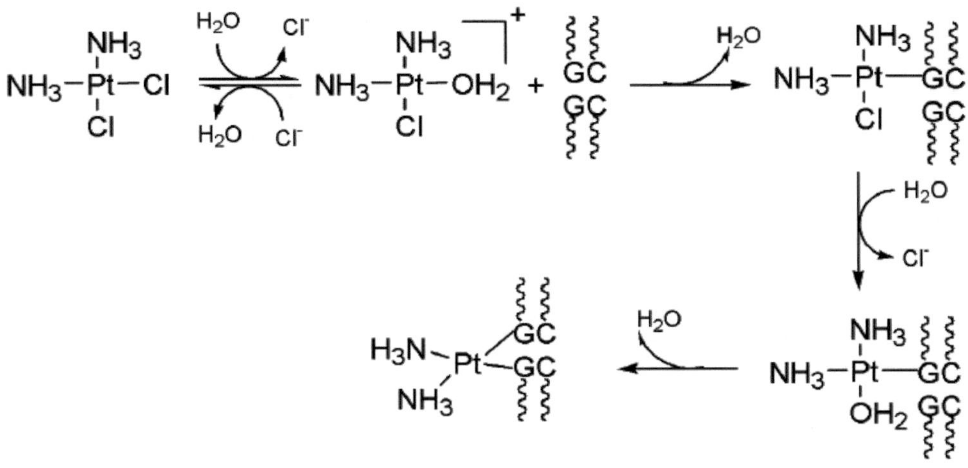

Figure 7.
Assumed mechanism for the formation of cisplatin-DNA adducts.

Cisplatin-DNA sequence selectivity has been given great attention from the research community. Several studies show that cisplatin first binds monofunctionally to guanine N7 and is particularly reactive towards Gn-runs ($n \geq 2$) (**Figure 7**) [18, 20, 21]. The high nucleophilicity of Gn-runs attracts the positively charged cisplatin monoaqua specie. The lifetime of the monoadduct is relatively long and it has therefore been suggested that the initial monoadduct is crucial for the type of cross-linked adduct formed and thus for the cytotoxic properties of the Pt complex. The main factors influencing the mono-functional binding affinity in DNA are thought to be [18] the type of bound nucleotide and of the adjacent residues, the steric effects of the Pt complex, the hydrogen binding properties of the Pt-DNA adduct and the DNA conformation.

8. Sequence specific structural perturbation

The formation of a cisplatin adduct with the GpG bases requires a significant tilting of the bases leading to a perturbation of the regular B-DNA conformation. The structural perturbation has been shown to be specifically recognized by a number of cellular proteins, including proteins with high-mobility group (HMG) binding domains and the TATA box binding protein [22]. It is believed that (some of) these recognition proteins mediate the cellular response which finally induces cell death by apoptosis or necrosis. In some cases, relatively subtle changes in the adduct structure can affect the recognition and the biological effects in a major way. This is exemplified by the cisplatin analogue oxaliplatin which forms similar G*G*-Pt adducts as cisplatin [18]. However, the oxaliplatin-G*G adducts differ in repair efficiency, mutagenesis and translesion synthesis, believed to be related to the differential activity of the two drugs (oxaliplatin is used, in combination with 5-fluorouracil, for the treatment of colorectal cancers against which cisplatin is inactive). The evaluation of the structural details of the platinum-DNA adducts and of their effects on protein, recognition can therefore help to understand why the biological activities of two similar platinum compounds (e.g., cisplatin versus oxaliplatin) are different. So far only nine cisplatin-DNA adducts have been characterized by NMR and/or x-ray crystallography. These structures were extensively reviewed by Ano et al. and found to be basically similar in structure. The cisplatin-GG adduct kinks the double helix approx. Approximately 60 towards the major groove and induces N sugar pucker for X of 5′ XG*, 5′ G* and the C complementary to 3′ G* [18].

This metal-based compound or coordination compounds that bind to DNA have been an active area of research since the discovery of cisplatin and the platinum-based drugs. The transition-metal compounds bind to DNA through several ways and different factors that promote it, such as the intercalant ligand and the nature and position of the substituent over it. Several techniques to follow metal-based drugs interactions with DNA are used as a powerful tool in order to reach a deep knowledge of the parameters involved in the stabilization of coordination compound-DNA adduct.

9. Methods for elucidation of DNA-anticancer drug interactions

DNA damaging agents (drugs that interfere with DNA function by chemically modifying specific nucleotides) includes mitomycin-C and echinomycin.

9.1 Mitomycin-C

Mitomycin-C is a well-characterized antitumor antibiotic that forms a covalent interaction with DNA after reductive activation (**Figure 8**). The activated antibiotic forms a cross-linking structure between guanine bases on adjacent strands of DNA therefore inhibiting single strand formation [8].

9.2 Echinomycin

Several studies have proved that both echinomycin quinoxaline rings bisintercalate into DNA, with CG selectivity, while the inner part of the depsipeptide establishes H-bonds with the DNA bases of the minor groove region of the two base pairs comprised between the chromophores (**Figure 9**) [8].

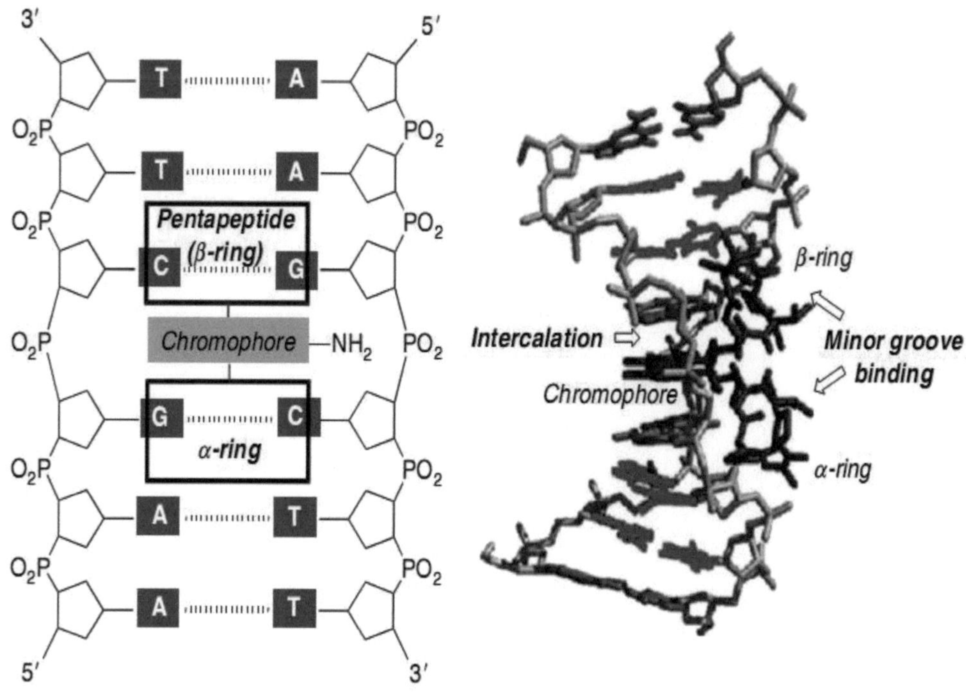

Figure 8.
Schematic interaction between DNA and mitomycin-C.

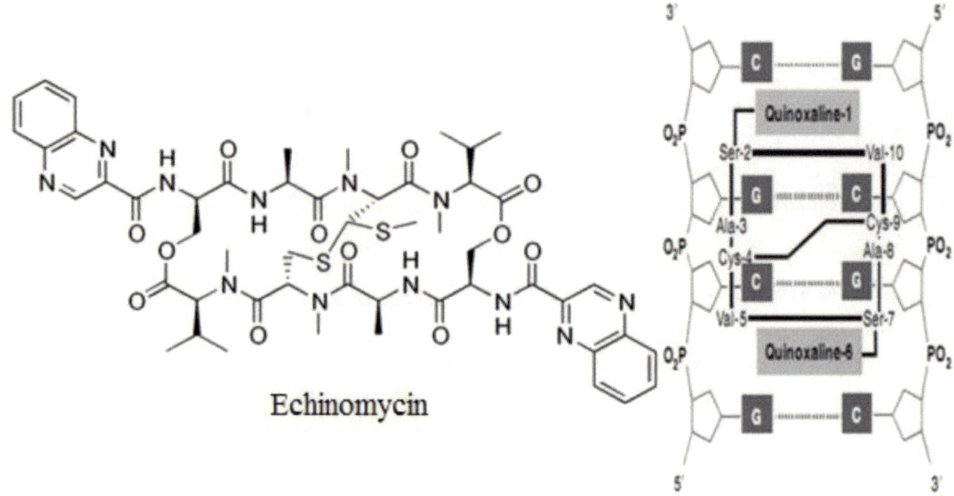

Figure 9.
Schematic interaction between DNA and echinomycin.

10. Mechanisms of anticancer drug-DNA interaction

The addition of anticancer drugs to a DNA molecule creates a new bond. Some examples for these mechanisms include intercalating agents, intercalating reagents (II), and bleomycins.

10.1 Intercalating agent

This agent contains planar aromatic or heteroaromatic ring systems (dactinomycin as an example), binding to sugar phosphate backbone by cyclic peptide or by NH_3. The planar systems slip between the layers of nucleic acid pairs and disrupt the shape of the helix. The preference is often shown for the minor or major groove. The intercalation prevents replication and transcription. In addition, the intercalation inhibits

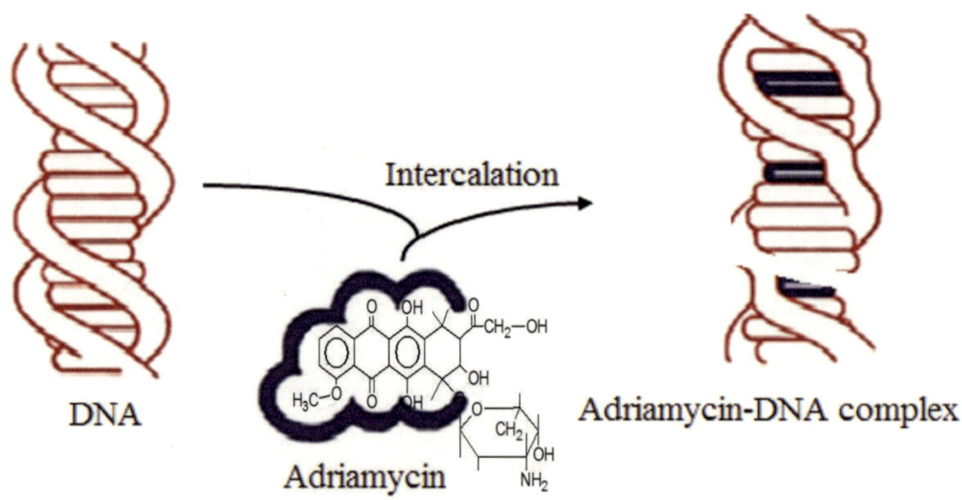

Figure 10.
Diagrammatic model illustrating intercalation of the flat part of the molecule of adriamycin (in black) into DNA, presenting the local unwinding of the helical structure.

topoisomerase II (an enzyme that relieves the strain in the DNA helix by temporarily cleaving the DNA chain and crossing an intact strand through the broken strand). Another example is the intercalation of the flat part of the molecule of Adriamycin into DNA, presenting the local unwinding of the helical structure (**Figure 10**) [23].

10.2 Intercalating reagents (II)

During replication, supercoiled DNA is unwound by the helicase. The thereby created tension is removed by the topoisomerase II (topo II) that cuts and rejoins the DNA strands. When doxorubicin is bound to the DNA it stabilizes the DNA-topo (II) complex at the point where the enzyme is covalently bound (**Figure 11**) [1, 24].

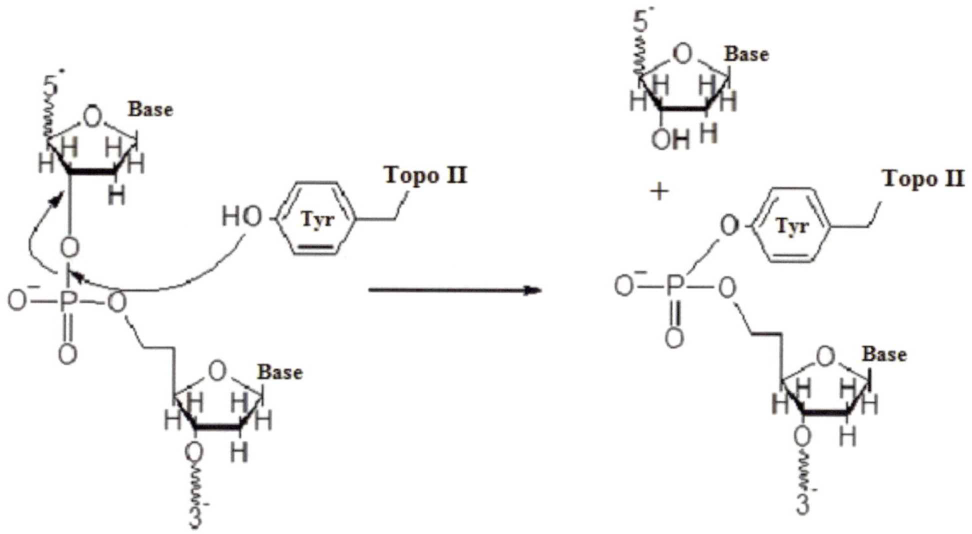

Figure 11.
Stabilizations of DNA-topo (II) complex.

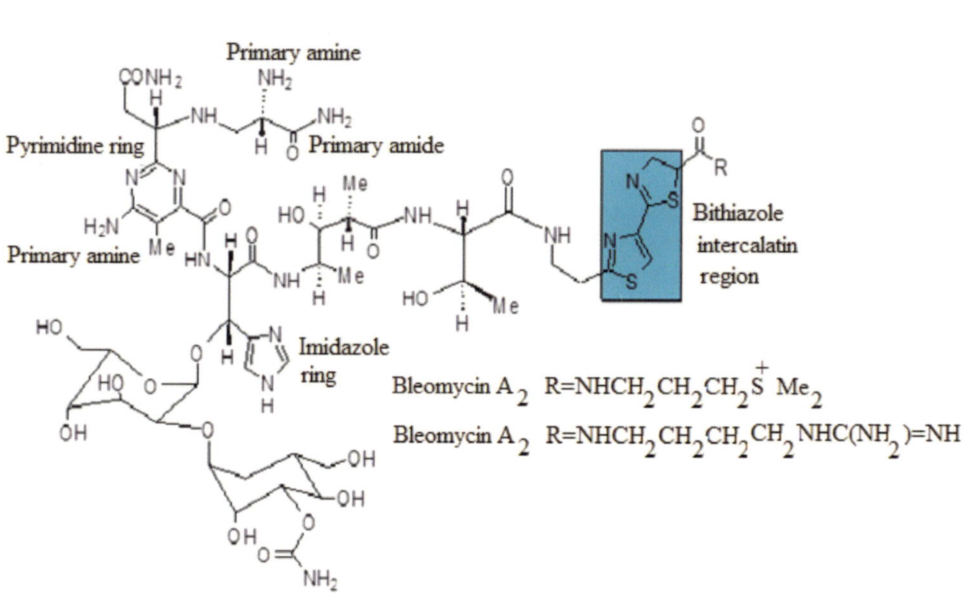

Figure 12.
The intercalating region (in blue color) of bleomycin A2 via the bithiazole moiety to DNA.

10.3 Bleomycin A_2

The bleomycin A_2 intercalate via the bithiazole moiety (DNA-binding domain) (**Figure 12**). The bithiazole moiety intercalates into the double helix and the attached side chain containing a sulfonium ion is attracted to the phosphodiester backbone. In addition, the N-atoms of the primary amines, pyrimidine ring and imidazole ring chelate Fe, which is involved in the formation of superoxide radicals, which subsequently act to cut DNA between purine and pyrimidine nucleotides [25].

11. Techniques for studying drug-DNA interactions

Various analytical techniques have been used for studying drug-DNA interactions (interaction between DNA and small ligand molecules that are potentially of pharmaceutical importance). Several instrumental techniques (emission and absorption spectroscopic) such as infrared (IR), UV-visible, nuclear magnetic resonance (NMR) spectroscopies, circular dichroism, atomic force microscopy (AFM), electrophoresis, mass spectrometry, viscosity measurements (viscometry), UV thermal denaturation studies, and cyclic, square wave and differential pulse voltammetry, etc., were used to study such interactions. These techniques have been used as a major tool to characterize the nature of drug-DNA complexation and the effects of such interaction on the structure of DNA. In addition, these techniques are regularly applied to monitor interactions of drugs with DNA because these optical properties are easily measured and tend to be quite sensitive to the environment. Moreover, these techniques provide various types of information (qualitative or quantitative) and at the same time complement each other to provide full picture of drug-DNA interaction and aid in the development of new drugs. In addition, the information gained from this part might be useful for the development of potential survey for DNA structure and new therapeutic reagents for tumors and other diseases. In this part of the chapter, we will focus on FT-IR, UV-Visible, NMR, AFM and viscosity measurements [5].

11.1 Fourier transform infrared spectroscopy

Fourier transform infrared (FT-IR) spectroscopy is a widely used technique to study interactions of nucleic acids (DNA and RNA) and proteins with anticancer drugs and other cytotoxic agents in solutions [26, 27]. In addition, it can generate structural information of the whole molecule in a single spectrum as a photograph of all conformations present in the sample that can distinguish among A-, B- and Z-forms of DNA, triple stranded helices, and other structural patterns. In addition, it is a powerful tool to study interactions of DNA with drugs and the effects of such interactions in the structure of DNA, and providing some insights about the mechanism of drug action. The technique is ideal for systematic studies of nucleic acids (e.g., sequence variations, covalent modifications), since it is fast, nondestructive, and requires only small amount of sample [28].

IR spectrum can be divided into four characteristic spectral ranges. The region between 1800 and 1550 cm^{-1} corresponds to the in-plane double bond vibrations of the nucleic bases (C═O, C═N, C═C and N—H bending vibrations of bases). These bands are sensitive to changes in the base stacking and base pairing interactions. Bands occurring in the interval 1500–1250 cm^{-1} assigned to vibrations of the bases and base-sugar connections are strongly related to the conformational changes of the backbone chain and glycosidic bond rotation. The range 1250–1000 cm^{-1} involves sugar phosphate vibrations, such as, PO_2 symmetric and asymmetric stretching vibrations and C—O stretching vibrations. These vibrations show high sensitivity to conformational changes in the backbone. The range 1000–800 cm^{-1} is characteristic for bands associated with vibrations of sugars which correlate with the various nucleic acid sugar puckering modes (C2'-endo and C3'-endo) [29, 30].

Due to interfering absorption bands of water at 1650 cm^{-1} and below 950 cm^{-1}, spectra are generally recorded also in D_2O, where these bands move to 1200 cm^{-1}, and below 750 cm^{-1}. Combination of results from both spectra allows obtaining a complete spectrum. The use of D_2O also causes shifts in nucleic acid absorptions, resulting from deuterium exchange of labile NH protons, and these can be used to monitor H–D exchange processes. A method to remove water signals in the spectra is water subtraction, using a sodium chloride (NaCl) solution as reference. D_2O is used to allow shifts in the absorption of nucleic acid in order to monitor H–D exchange processes. Four regions, each having marker bands showing either nucleic acid interactions or conformations, are presented in **Figure 13** [31, 32].

The ring vibrations of nitrogenous bases (C═O, C═N stretching), PO_2 stretching vibrations (symmetric and asymmetric) and deoxyribose stretching of DNA backbone are confined in the spectral region between 1800 and 700 cm^{-1}. The vibrational bands of DNA at 1710, 1662, 1613 and 1492 cm^{-1} are assigned to guanine (G), thymine (T), adenine (A) and cytosine (C) nitrogenous bases, respectively. Bands at 1228 and 1087 cm^{-1} denote phosphate asymmetric and symmetric vibrations, respectively. These are the prominent bands of pure DNA, which are monitored during carboplatin-DNA interaction at different ratios. Changes in these bands are shown in **Figure 14** [33]. After carboplatin addition to DNA solution, G-band at 1710 shifts to 1702–3, T-band at 1662 shifts to 1655 and A-band at 1613 shifts towards lower wave number 1609–10 cm^{-1}. These shifting can be attributed to direct platin binding to G (N7), T (O_2), and A (N7) of DNA bases. No major shifting is observed for phosphate asymmetric and symmetric vibrations indicating no external binding. The plots of the relative intensity (*R i*) of several peaks of DNA in-plane vibrations related to A–T, G–C base pairs and the PO_2^- stretching vibrations such as 1717 (G), 1663 (T), 1609 (A), 1492 (C), and 1222 cm^{-1} (PO_2^- groups), against the compound concentrations can be obtained after peak normalization using formula (1) [5, 34]:

$$R_i = \frac{I_i}{l_{968}} \tag{1}$$

where R_i is the relative intensity, I_i is the intensity of absorption peak for pure DNA and DNA in the complex with *i* concentration of compound, and l_{968} is the intensity of the 968 cm^{-1} peak (internal reference) [35].

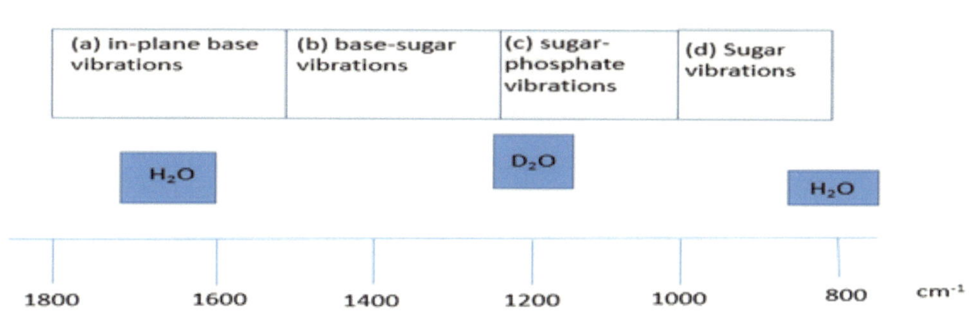

Figure 13.
The characteristics IR bands of DNA and aqueous solvents. (a) 1800–1500 cm^{-1} region is sensitive to effects of base pairing and base stacking; (b) 1500–1250 cm^{-1} region is sensitive to glycosidic bond rotation, backbone conformation, and sugar pucker; (c) 1250–1000 cm^{-1} region is sensitive to backbone conformation; and (d) 1000–800 cm_{-1} region is sensitive to sugar conformation.

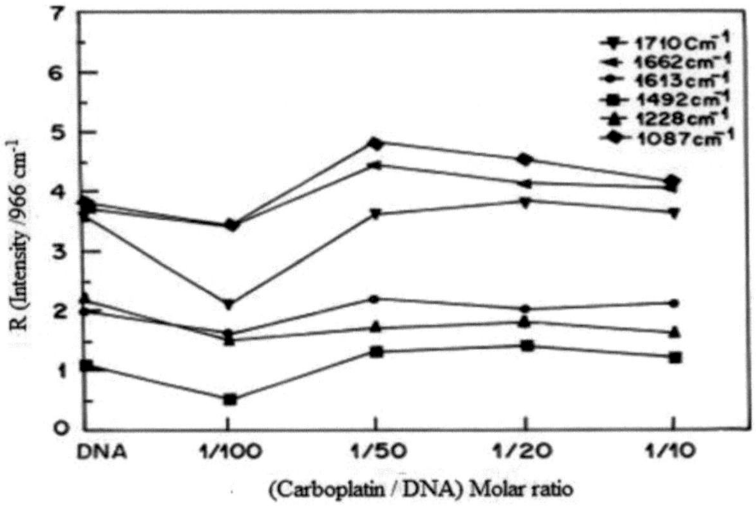

Figure 14.
Intensity ratio variations for DNA as a function of different carboplatin/DNA molar ratios.

Similarly, Raman spectroscopy, which also depends on characteristic group vibrational frequencies, can be used together with infrared spectra to study vibrations in DNA. It is useful because Raman and IR spectroscopy provide complementary information.

11.2 UV-visible spectroscopy

UV-visible absorption spectroscopy can be utilized to detect the DNA-ligand interaction by measuring the changes in the absorption properties of the DNA molecules or the ligand. The UV-vis absorption spectrum of DNA displays a broad band in the range of 200–350 nm in the UV region, with a maximum situated at 260 nm. The maximum is due to the chromophoric groups in pyrimidine and purine moieties responsible for the electronic transitions. The utilization of this simple and versatile technique enables an accurate estimation of the DNA molar concentration based on absorbance measurement at 260 nm. To measure the interaction between ligands and DNA, a hypochromic shift is utilized because the monitoring of the values of absorbance enables studying of the melting action of DNA. Apart from versatility, other major advantages of UV-vis absorption spectroscopy include simplicity, reproducibility, and good sensitivity [36, 37].

11.3 Nuclear magnetic resonance spectroscopy

Binding between ligands and the molecules of DNA causes a significant change in the chemical shift of the values presented in **Table 2** [32]. For example, applying thermal denaturing in order to un-stack the base-pair double-helical DNA to form two ss-DNAs is often accompanied by the ^{1}H resonances' downfield shift for non-exchangeable protons.

The broadening of ^{1}H NMR resonances of DNA upon addition of an appropriate minor groove binding compound is one type of evidence of complex formation in DNA ^{31}P-NMR spectroscopy has also been used to provide important information concerning the binding of intercalators to DNA. The ^{31}P chemical shifts are sensitive DNA conformational changes, and hence intercalating drugs cause downfield shift, while divalent cations causes up field shifts in the ^{31}P signal [38].

Proton type	Expected chemical shift[a] (ppm)	Proton type	Expected chemical shift[a] (ppm)
T5 (CH_3)	1.00–2.00	A 2 (CH); A 8 (CH); G 8 (CH) T 6 (CH) C 6 (CH)	6.50–8.20
Sugar 2′ (CH_2)	2.00–3.00		
Sugar 5′ terminal (CH_2)	3.70		
Sugar 5′ (CH_2); 4'(CH)	4.00–4.50	C 4 (NH_2) (H-1)[b]	6.40–6.80
Sugar 3′ (CH)	4.50–5.20	C 4 (NH_2) (H-2)[b]	8.30–8.50 ppm
Sugar 1′ (CH)	5.30–6.20	G 1 (NH)	12.50–13.00 ppm
C 5 (CH)	5.30–6.20	T 3 (NH)	13.50–14.00 ppm

[a] *Chemical shifts relative to internal TSP (3-(trimethylsilyl)propionic acid).*
[b] *For Watson-Crick base pairs (CG).*

Table 2.
Typical ranges of chemical shifts for ^1H NMR spectra of nucleic acids.

11.4 Mass spectrometry

Mass spectrometry (MS) has become one of the most common techniques adopted to study interactions between DNA and small ligand molecules. The ability of mass spectrometry to investigate drug-DNA interactions have been reviewed recently. The binding stoichiometry, the relative binding affinities and the binding constants for DNA double helices of various sequences may be determined. Electrospray ionization (ESI) is the most common ionization method used in the study of biomolecules due to its soft ionization. Using ESI techniques, biomolecules can be transferred from the solution to the mass spectrometer with minimal fragmentation and, so, both the mass of the DNA and the mass of the DNA-ligand complex can be determined, as the non-covalent interactions that formed the complex are not altered during the electrospray process [39–41]. Focusing on the use of ESI-MS to study complexes, MS gives a signal for each species with a different mass and so it is very straightforward to establish the stoichiometry of the complexes. ESI-MS signals enable several calculations to be performed. The number of DNA strands involved, the number of bound cations (if present) and the number of bound ligands, among others. Taking into account the structure of the nucleic acids, ESI-MS studies are performed using negative polarity. It is well known that the phosphodiester backbone of DNA is fully deprotonated under usual working conditions. In general, in order to preserve their structure, nucleic acid solutions are prepared with monovalent ions. Perylene derivatives, such as, N,N-bis-(2-(dimethylamino)ethyl)-3,4,9,10-perylenetetracarboxylic acid diimide, favor π-π interactions with the G-tetrad surface. Moreover, 5,10,15,20-tetrakis-(1-methyl-4-pyridyl)-21H,23H-porphine is an effective telomerase inhibitor, also binds to the G-quadruplex in the c-myc promoter [42].

11.5 Atomic force microscopy

Atomic force microscopy (AFM) can be used to distinguish proteins bound to nucleic acid templates. One of the great advantages of the atomic force microscope, particularly with respect to the imaging of biological specimens, is that it can work in fluid, so that experiments can be performed under near physiological conditions and allowing the imaging of interactions and transactions between molecules in real time [43]. AFM techniques will play a larger role in studying interactions between biological specimens, such as ligand-receptor and protein-DNA systems, and can be applied to the study of drug interactions with a variety of biological specimens [5].

Drug-DNA complexes have been studied with AFM to determine the binding force between them. This is of considerable interest since nucleic acid ligands are commonly used as anticancer drugs and in the treatment of genetic diseases. However, determining whether they bind to DNA by intercalation within major and/or minor grooves, by normal modes, or by a combination of these modes can often be difficult. AFM was used to study drug binding mode, affinity, and exclusion number by comparing the length of DNA fragments that have and have not been exposed to the drug. It is well known that if intercalative binding is occurring, the DNA strand increases in length. Moreover, the degree of lengthening is informative in determining the binding affinity and the site-exclusion number. AFM was shown to be an effective means of seeing and measuring any changes in the DNA strand. For example, when it exposed to ethidium, the DNA strand was shown through AFM to have increased in length from 3300 to 5250 nm, this indicating the intercalative mode of binding. Similarly, AFM intercalative binding studies showed the increase in the DNA strand, from 3300 to 4670 nm, upon exposure to daunomycin. This technique has also successfully been applied to new drugs in which the mode of binding was unclear. For example, exposure of 2,5-bis(4-amidinophenyl) (APF), did not produce lengthening of the DNA strands, indicating that the drug binds by non-intercalative modes. The different structural changes and binding processes of the DNA occur because of interactions with these two components [5].

11.6 Viscosity measurements

DNA viscosity is sensitive to DNA length change, for this reason, its measurement upon the addition of a compound is often concerned as the least ambiguous and most critical method to clarify the interaction mode of a compound with DNA and this will provide reliable evidence for the intercalative binding mode. Relative viscosity measurements have proved to be a reliable method for the assignment of the mode of binding compounds to DNA. In the case of classical intercalation, DNA base pairs are separated in order to host the bound compound resulting in the lengthening of the DNA helix and subsequently increased DNA viscosity. On the other side, the binding of a compound exclusively in DNA grooves by means of partial and/or non-classic intercalation, under same conditions, causes a bend or kink in the DNA helix and reducing its effective length and, as a result, DNA solution viscosity is decreased, or it remains unchanged.

Figure 15 show the interaction of three Schiff base compounds of N′-substituted benzohydrazide and sulfonohydrazide derivatives: (1) N′-(2-hydroxy-3-methoxybenzylidene)-4-tert-butylbenzohydrazide, (2) N′-(5-bromo-2 hydroxy-benzylidene)-4-tert-butylbenzohydrazide and (3) N′-(2-hydroxy-3-methoxy-benzylide-ne)-4-methylbenzenesulfonohydrazide with SS-DNA [44]. This can be explained by the insertion of the compounds in between the DNA base pairs, leading to an increase in the separation of base pairs at intercalation sites and, thus, an increase in DNA length [45].

The viscosity data show that there are at least two phases of binding between the complex and CT-DNA. At lower concentration of the complex, the viscosity first decreases and then increases at higher concentration of complex. This slow increase in viscosity is an indication of groove binding [11].

Figure 16 indicate that with increasing amount of (3-(3,5 dimethyl-phenylimino) methyl)benzene-1,2-diol (HL), the relative viscosity of DNA first remains constant and then increases [46]. This observation supports that HL bind through intercalation mode but with different affinity, i.e., also show some affinity for binding with grooves of DNA through hydrogen bonding, typically to N_3 of adenine and O_2 of thymine. However, strong binding is presumably due to intercalation with DNA [11].

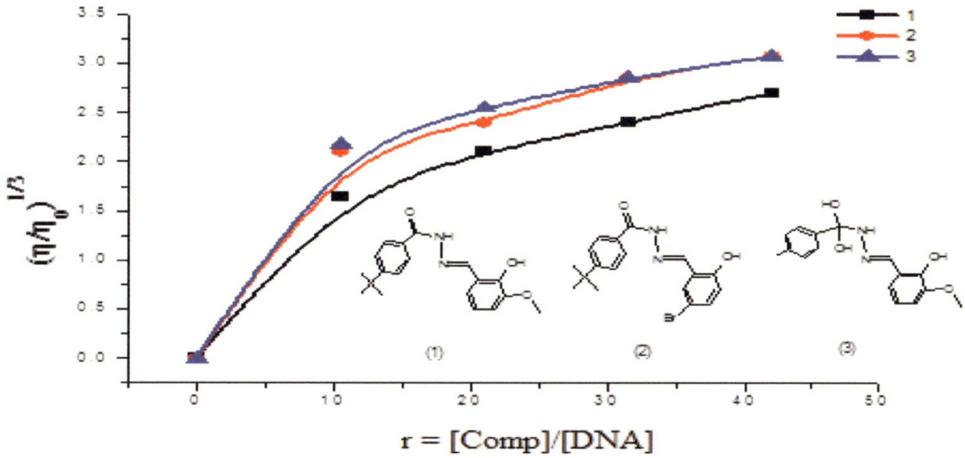

Figure 15.
Effects of increasing amount of compounds (1–3) on relative viscosity of SS-DNA at 25 ± 0.1°C. [DNA] = 7.2 μM, r = 0, 6.9, 13.9, 20.8, 27.8.

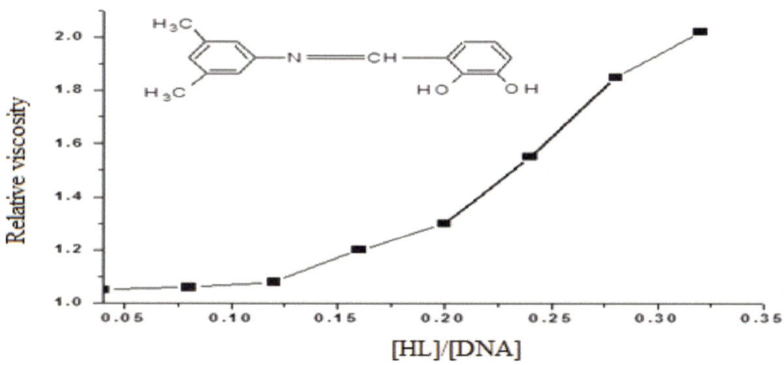

Figure 16.
Effects of increasing amount of HL on relative viscosity of CT-DNA at 25 ± 0.1°C. [DNA] = 2.37 × 10^{-5} M.

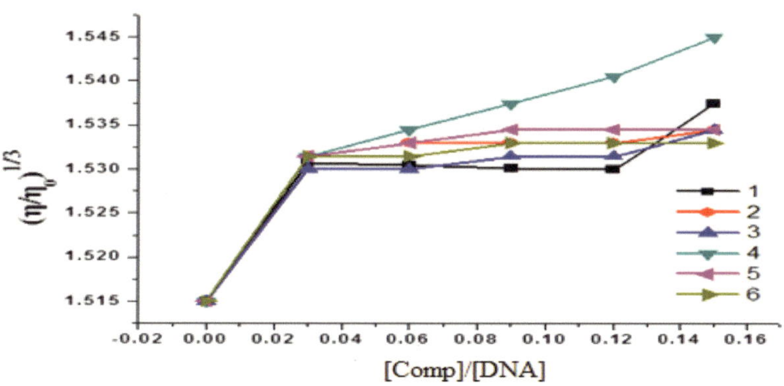

Figure 17.
(1) Effect of increasing amount of the complexes [Ni(hhmh)$_2$], (2) [Ni(bhmh)$_2$], (3) [Ni(ihmh)$_2$], (4) [Ni(PPh$_3$)(hpeh)], (5) [Ni(PPh$_3$)(bpeh)] and (6) [Ni(PPh$_3$)(ipeh)] on the relative viscosity of HS-DNA at 16(±0.L)°C.

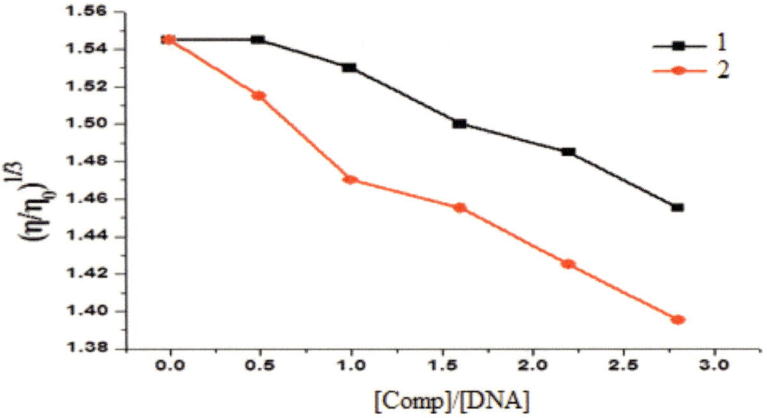

Figure 18.
(1) Effects of increasing amount of tri-n-butyltin (IV) 3-[(3',5'dimethylphenylamino)] propanoate and (2) triphenyltin(IV) 3-[(3',5'dimethylphenylamino)]propanoate on relative viscosity of SS-DNA at 25 ± 0.1°C, [DNA] = 1.86 × 10^{-4} M.

Figure 17 [47] and **Figure 18** [48] shows the electrostatic binding mode of nickel and organotin(IV) complexes with DNA, respectively. The viscosity of DNA remains essentially unchanged on the addition of the nickel complexes while it decreases in case of organotin(IV) complexes [11].

12. Conclusions

This chapter has focused on drug-DNA interactions and their study by various analytical techniques such as IR spectroscopy, viscosity measurements, MS and AFM. These techniques are used to evaluate the binding mode as well as binding strength of the complex formed between drug and DNA. The study should be useful for the development of potential survey for DNA structure and new therapeutic reagents for tumors and other diseases. Fundamentally, drugs interact with DNA through two different ways, covalent and/or non-covalent modes. Covalent binders act as alkylating agents as they alkylate the nucleotides of DNA, while, the non-covalent binders interact by three different ways: (i) intercalation, (ii) groove binding, and (iii) external binding (on the outside of the helix). Different spectroscopic techniques are generally, powerful tools to study interactions of DNA with drugs and the effects of such interactions in the structure of DNA, providing some insights about the mechanism of drug action. The binding stoichiometry, the relative binding affinities and the binding constants for DNA double helices of various sequences.

Author details

Saad Hmoud Alotaibi[1] and Awad Abdalla Momen[1,2*]

1 Department of Chemistry, Turabah University College, Taif University, Saudi Arabia

2 Department of Chemistry, College of Industrial and Applied Science, University of Bahri, Sudan

*Address all correspondence to: aamomena@yahoo.com

References

[1] Wang JC. Nature reviews. Cellular roles of DNA topoisomerases: A molecular perspective. Molecular and Cellular Biology. 2002;**3**:430-440

[2] Guerra CF, Bickelhaupt FM. Watson-Crick hydrogen bonds: Nature and role in DNA replication. Modern Methods for Theoretical Physical Chemistry of Biopolymers. 2006;**19**:79-97

[3] Langkjæra N, Wengela J, Pasternak A. Watson–Crick hydrogen bonding of unlocked nucleic acids. Bioorganic & Medicinal Chemistry Letters. 2015;**25**(22):5064-5066

[4] García-Ramos J, Galindo-Murillo R, Cortés-Guzmán F, Ruiz-Azuara L. Metal-based drug-DNA interactions. Journal of the Mexican Chemical Society. 2013;**57**(3):245-259

[5] Sirajuddin M, Haider A, Ali S. Analytical techniques for the study of drug-DNA interactions. International Journal of Advanced Research. 2013;**9**:510-530

[6] Kumar S, Pandya P, Pandav K, Gupta S, Chopra A. Structural studies on ligand—DNA system. A rubst approach in drug design. Journal of Biosciences. 2012;**37**:553-561

[7] Palchaudhuri R, Hergenrother P. DNA as a target for anticancer compounds: Methods to determine the mode of binding and the mechanism of action. Current Opinion in Biotechnology. 2007;**18**:497-503

[8] Goftar M, Kor N, Kor Z. DNA intercalators and using them as anticancer drugs. International Journal of Advanced Biological and Biomedical Research. 2014;**2**(3):811-822

[9] Hajian R, Shams N, Mohagheghian M. Study on the interaction between doxorubicin and deoxyribonucleic acid with the use of methylene blue as a probe. Journal of the Brazilian Chemical Society. 2009;**20**(8):1399-1405

[10] Sechi M, Derudas M, Dallocchio R, Dessì A, Cosseddu A, Paglietti G. DNA binders: Evaluation of DNA-interactive ability, design, and synthesis of novel intercalating agents. Letters in Drug Design & Discovery. 2009;**6**:56-62

[11] Sirajuddin M, Ali S, Badshah A. Drug–DNA interactions and their study by UV–visible, fluorescence spectroscopies and cyclic voltametry. Journal of Photochemistry and Photobiology B: Biology. 2013;**124**:1-19

[12] Paul A, Bhattacharya S. Chemistry and biology of DNA-binding small molecules. Current Science. 2012;**102**(2):212-231

[13] Pilch D, Poklar N, Gelfand C, Law S, Breslauer K, Baird E, et al. Binding of a hairpin polyamide in the minor groove of DNA: Sequence-specific enthalpic discrimination. Proceedings of the National Academy of Sciences of the United States of America. 1996;**93**:8306-8311

[14] Mrksich M, Parks ME, Dervan PB. Hairpin peptide motif. A new class of oligopeptides for sequence-specific recognition in the minor groove of double-helical DNA. Journal of the American Chemical Society. 1994;**116**:7983-7988

[15] Vollmer M, Nägele E, Hörth P. Differential proteome analysis: Two-dimensional nano–LC/MS of *E. coli* proteome grown on different carbon sources. Journal of Biomolecular Technology. 2003;**21**:289-307

[16] Mandal S, Moudgil M, Mandal S. Rational drug design. European Journal of Pharmacology. 2009;**625**(1-3):90-100

[17] Newlands ES, Stevens MFG, Wedge SR, Wheelhouse RT, Brock C. Temozolomide: A review of its discovery, chemical properties, pre-clinical development and clinical trials. Cancer Treatment Reviews. 1997;**23**:35-61

[18] Skauge T. Antibacterial and anticancer drugs—Interaction with DNA (i) antibacterial fluoroquinolones (ii) anticancer cis–platinum(II) complexes [PhD thesis]. Department of Chemistry, Faculty of Science, University of Bergen; 2006

[19] Eastman A, Jennerwein MM, Nagel DL. Characterization of bifunctional adducts produced in DNA by trans–diamminedichloroplatinum(II). Chemico-Biological Interactions. 1988;**67**:71-80

[20] Davies M, Berners-Price SJ, Hambley TW. Slowing of cisplatin aquation in the presence of DNA but not in the presence of phosphate: Improved understanding of sequence selectivity and the roles of monoaquated and diaquated species in the binding of cisplatin to DNA. Inorganic Chemistry. 2000;**39**:5603-5613

[21] Legendre F, Bas V, Kozelka J, Chottard J. A complete kinetic study of GG versus AG platination suggests that the doubly aquated derivatives of cisplatin are the actual DNA binding species. Chemistry—A European Journal. 2000;**6**:2002-2010

[22] Zamble D, Lippard SJ. The response of cellular proteins to cisplatin-damaged DNA. In: Lippert B, editor. Cisplatin. Chemistry and Biochemistry of a Leading Anticancer Drug. Zürich, Weinheim: Verlag Helvetica Chimica Acta, Wiley-VCH; 1999. pp. 73-110

[23] Krogsgaard-Larsen P. Textbook of Drug Design and Discovery. 3rd ed. London, U.K: Taylor and Francis; 2002

[24] Champoux J. DNA Topoisomerase: Structure, function and mechanism. Annual Review of Biochemistry. 2001;**70**:369-413

[25] Hecht S. Bleomycin: New perspectives on the mechanism of action. Journal of Natural Products. 2000;**63**(1):158-168

[26] Froehlich E, Gupta A, Provencher-Mandeville J, Asselin E, Bariyanga J, Berube G, et al. Study of DNA interactions with steroidal and nonsteroidal estrogen-platinum (II)-based anticancer drugs. DNA and Cell Biology. 2009;**28**(1):31-39

[27] Nafisi S, Sobhanmanesh A, Alimoghaddam K, Ghavamzadeh A, Tajmir-Riahi HA. Interaction of arsenic trioxide As_2O_3 with DNA and RNA. DNA and Cell Biology. 2005;**24**(10):634-640

[28] Charak S, Jangir DK, Tyagi G, Mehrotra R. Interaction studies of epirubicin with DNA using spectroscopic techniques. Journal of Molecular Structure. 2011;**1000**:150-154

[29] Lee S, Debenedetti P, Errington J, Pethica B, Moore D. A calorimetric and spectroscopic study of DNA at low hydration. The Journal of Physical Chemistry. B. 2004;**108**(9):3098-3106

[30] Pevsner A, Diem M. IR spectroscopic studies of major cellular components. III. Hydration of protein, nucleic acid, and phospholipid films. Biopolymers. 2003;**72**(4):282-289

[31] Banyay M, Sarkar M, Graslund A. A library of IR bands of nucleic acids in solution. Biophysical Chemistry. 2003;**104**(2):477-488

[32] González-Ruiz V, Olives A, Martín M, Ribelles P, Ramos M, Menéndez J. An overview of analytical techniques employed to evidence drug–DNA interactions. Applications to the

design of genosensors. In: Biomedical Engineering, Trends, Research and Technologies. 2011. pp. 65-90

[33] Nafisi S, Saboury AA, Keramat N, Neault JF, Tajmir-Riahi HA. Stability and structural features of DNA intercalation with ethidium bromide, acridine orange and methylene blue. Journal of Molecular Structure. 2007;**827**(1-3):35-43

[34] Nafisi S, Bonsaii M, Alexis V, Glick J. Binding of 2–acetylaminofluorene to DNA. DNA and Cell Biology. 2011;**30**(11):955-962

[35] Andrushchenko V., Leonenko Z., Cramb D., van de Sande H, Wieser H. VCD and AFM study of DNA interaction with Cr^{3+} ions, VCD and AFM evidence of DNA condensation. Biopolymers, 2002;(61):243-260

[36] Kumar K, Reddy K, Satyanarayana S. Synthesis, DNA interaction and photocleavage studies of ruthenium(II) complexes with 2-(pyrrole) imidazo[4,5-*f*]-1,10–phenanthroline as an intercalative ligand. Transition Metal Chemistry. 2010;**35**:713-720

[37] Tan C, Liu J, Chen L, Shi S, Li L. Synthesis, structural characteristics, DNA binding properties and cytotoxicity studies of a series of Ru(III) complexes. Journal of Inorganic Biochemistry. 2008;**102**:1644-1653

[38] Alotaibi S. Novel spectroscopic tools to differentiate molecule–DNA binding interactions, sense DNA and track DNA melting [PhD dissertations]. Western Michigan University; 2015

[39] Brodbelt J. Evaluation of DNA/ligand interactions by electrospray ionization mass spectrometry. Annuals Reviews in Analytical Chemistry. 2010;**3**:67-87

[40] Jaumot J, Gargallo R. Experimental methods for studying the interactions between G–quadruplex structures and ligands. Current Pharmaceutical Design. 2012;**18**(14):1900-1916

[41] Rosu F, De Pauw E, Gabelica V. Electrospray mass spectrometry to study drug–nucleic acids interactions. Biochimie. 2008;**90**:1074-1087

[42] Li H, Liu Y, Lin S, Yuan G. Spectroscopy probing of the formation, recognition, and conversion of a G–quadruplex in the promoter region of the bcl-2 oncogene chemistry. A European Journal. 2009;**15**(10):2445-2452

[43] Edwardson J, Henderson RM. Atomic force microscopy and drug discovery. Research Focus/Reviews. 2004;**9**(2):64-71

[44] Sirajuddin M, Uddin N, Ali S, Tahir MN. Potential bioactive Schiff base compounds: synthesis, characterization, X-ray structures, biological screenings and interaction with Salmon sperm DNA. Spectrochimica Acta Part A. 2013;**116**:111-121

[45] Dimiza F, Perdih F, Tangoulis V, Turel I, Kessissoglou D, Psomas G. Interaction of copper(II) with the non-steroidal anti-inflammatory drugs naproxen and diclofenac: Synthesis, structure, DNA- and albumin-binding. Journal of Inorganic Biochemistry. 2011;**105**(3):476-489

[46] Sirajuddin M, Ali S, Shah NA, Khan MR, Tahir MN. Synthesis, characterization, biological screenings and interaction with calf thymus DNA of a novel azomethine 3-((3,5-dimethylphenylimino)methyl) benzene-1,2-diol. Spectrochimica Acta. Part A, Molecular and Biomolecular Spectroscopy. 2012;**94**:134-142

[47] Chitrapriya N, Mahalingam V, Zeller M, Natarajan K. Synthesis, characterization, crystal structures and DNA binding studies of nickel(II)

hydrazone complexes. Inorganica Chimica Acta. 2010;**363**:3685-3693

[48] Shah F, Sirajuddin M, Ali S, Abbas SM, Tahir MN, Rizzoli C. Synthesis, spectroscopic characterization, X-ray structure and biological screenings of organotin (IV) 3-[(3,5-dichlorophenylamido)] propanoates. Inorganica Chimica Acta. 2013;(400):159-168